The OPL Optimization Programming Language

The OPL Optimization Programming Language

Pascal Van Hentenryck

With contributions by Irvin Lustig, Laurent Michel, and Jean-François Puget

The MIT Press
Cambridge, Massachusetts
London, England

© 1999 Massachusetts Institute of Technology

All rights reserved. No part of this book may be reproduced in any form by any electronic or mechanical means (including photocopying, recording, or information storage and retrieval) without permission in writing from the publisher.

This book was printed and bound in the United States of America.

Library of Congress Cataloging-in-Publication Data

Van Hentenryck, Pascal.
    The OPL optimization programming language / Pascal Van Hentenryck; with contributions by Irvin Lustig, Laurent Michel, and Jean-François Puget.
        p. cm.
    Includes bibliographical references and index.
    ISBN 0-262-72030-2 (pbk. : alk. paper)
    1. OPL (Computer program language) I. Lustig, Irvin. II. Title.
QA76.73.O215V37   1999
003′.35133–dc21   98-34698
                                                                                                                                                 CIP

Contents

	Preface	ix
1	**Introduction**	1
1.1	Background	1
1.2	OPL	6
1.3	Contents	8
1.4	Model Conventions and Disclaimers	9

I   THE LANGUAGE

2	**A Short Tour of OPL**	13
2.1	Linear and Integer Programming	13
2.2	Constraint Programming	31
2.3	Scheduling	50
2.4	Notes and References	53

3	**Models**	55
3.1	Syntactic Conventions	55
3.2	Terminal Symbols	55
3.3	Models	55

4	**Data Modeling**	57
4.1	Basic Data Types	57
4.2	Data Structures	59
4.3	Variables	68
4.4	Data Types for Scheduling Applications	70
4.5	Constraint Declarations	75
4.6	Data Consistency	76
4.7	Initialization	77

5 Expressions and Constraints — 85
5.1 Expressions and Relations — 85
5.2 Constraints — 97
5.3 Stating Constraints — 106

6 Formal Parameters — 111
6.1 Basic Formal Parameters — 111
6.2 Tuples of Parameters — 113
6.3 Filtering in Tuples of Parameters — 114
6.4 Modeling Issues — 115

7 Search — 121
7.1 The Try Instruction — 121
7.2 The Tryall Instruction — 124
7.3 Quantifiers — 125
7.4 Sequencing Choices — 127
7.5 Conditional Choices — 128
7.6 The While Instruction — 129
7.7 The Select Instruction — 129
7.8 The Let Statement — 130
7.9 The Once Statement — 130
7.10 Constraints — 132
7.11 Data-Driven Constructs — 132
7.12 Predefined Search Strategies — 134
7.13 Choices in Scheduling — 136

8 Display — 139
8.1 Displaying Data — 139
8.2 Filtering and Aggregating Results — 141
8.3 Computing Derived Results — 142

| 8.4 | Displaying Tuples | 142 |

II THE APPLICATION AREAS

9 Linear and Integer Programming — 145
9.1	Linear Programming	145
9.2	Integer Programming	157
9.3	Mixed Integer-Linear Programming	166
9.4	Piecewise Linear Programming	168
9.5	Notes and References	176

10 Constraint Programming — 177
10.1	Warehouse Location	177
10.2	Car Sequencing	184
10.3	The Euler Tour	191
10.4	Frequency Allocation	194
10.5	Rack Configuration	198
10.6	Notes and References	202

11 Scheduling — 203
11.1	Origin and Horizon	203
11.2	Activities	203
11.3	Unary Resources	205
11.4	Discrete Resources	218
11.5	Reservoirs	228
11.6	Alternative Resources	232
11.7	Notes and References	235

Bibliography 237

Index 241

Preface

Linear programming, integer programming, and combinatorial optimization problems are ubiquitous in many application areas such as planning, scheduling, sequencing, resource allocation, design, and configuration. Robust solvers are now available that solve large-scale linear programs and various classes of integer programs. However, many integer programming and combinatorial optimization problems are challenging from a computational standpoint: they are NP-complete or worse, and it is widely believed that no general and efficient algorithm exists for solving them. As a consequence, their solution requires considerable time and expertise in both the application domain (modeling) and algorithm design (solving). In addition, the resulting algorithmic solutions often involve substantial development effort, since the distance between the problem model and the computer algorithm may be large.

This book describes OPL (**O**ptimization **P**rogramming **L**anguage), a modeling language for combinatorial optimization that may simplify these optimization problems substantially. OPL was motivated by modeling languages such as AMPL and GAMS that provide computer equivalents to traditional algebraic notation. It provides similar support for modeling linear and integer programs and provides access to state-of-the-art linear programming algorithms. But OPL adds several new dimensions to modeling languages beyond the traditional support for linear and integer programming.

Perhaps the most significant new dimension of OPL is the support for constraint programming. The essence of constraint programming is a two-level architecture integrating a constraint and a programming component. The constraint component provides the basic operations of the architecture and consists of a system that reasons about fundamental properties of constraint systems such as the satisfiability and entailment of constraints. Operating around this level is a programming-language component that specifies how to combine the basic operations, often in nondeterministic ways, since algorithms for searching the solution space are so fundamental in combinatorial optimization. By supporting both the constraint and the programming components of the constraint programming architecture, OPL goes far beyond traditional modeling languages to let users specify search procedures tailored to the problem at hand. Another significant dimension added by OPL is its high-level support for scheduling and resource allocation applications, which are ubiquitous in industry. OPL provides novel modeling concepts such as activities and resources, and provides access to special-purpose algorithms such as the edge-finder procedure. Finally, OPL improves the expressiveness of traditional modeling languages by offering new concepts such as higher-order constraints, logical combinations of constraints, and many other new modeling tools.

Contents This book is a comprehensive presentation of OPL. It is not an introduction to combinatorial optimization or constraint programming. Readers should be familiar with combinatorial optimization, at least from an application standpoint. Part I describes the language itself; each chapter reviews some aspect of OPL in detail. These technical chapters are preceded by a short tour of OPL to introduce the main concepts and the computation model. Part II applies OPL in three application areas: linear and integer programming, constraint programming, and scheduling.

The book can be read in two ways. All readers should first take the short tour of OPL (Chapter 2). They can then read the book sequentially or, alternatively, go directly to Part II. There is some redundancy in the book to make this possible; in particular, some of the chapters in Part II contain a summary of relevant concepts when necessary.

OPL Studio OPL is the modeling language of OPL STUDIO, an integrated development environment for combinatorial optimization applications. In addition to standard editing and structuring capabilities, OPL STUDIO also contains functionalities for debugging and visualizing OPL statements. OPL STUDIO is rarely referred to in this book; the only notable exception is in Chapter 2, where the separation of the model and the data is discussed.

Industrial Version of OPL Studio An industrial implementation of OPL STUDIO is now available from ILOG under the name ILOG OPL STUDIO. Look at `http://www.ilog.com` for more information.

Acknowledgments I would like to express my gratitude to Irvin Lustig, Laurent Michel, and Jean-François Puget for numerous discussions on the design of OPL. Irvin and Laurent also proofread the entire manuscript in depth. Special thanks to Christiane Bracchi for proofreading and testing all the examples in this book, to Yves Deville for commenting on a first version of this book, to Katrina Avery and Gwyneth Owens Butera for proofreading the entire manuscript, and, as always, to Bob Prior whose patience is seemingly without limit. And, of course, this book would have been written much faster, but with only a fraction of the pleasure, without the increasingly creative, and frequent, interruptions of Anne, Thomas, and Maité.

Pascal Van Hentenryck
Barrington, R.I.
May, 1998

1 Introduction

Linear programming, integer programming, and combinatorial optimization problems arise in a variety of application areas, which includes planning, scheduling, sequencing, resource allocation, design, and configuration. In the last decades, robust solvers were developed and implemented to solve large scale linear programs and various classes of integer programs. However, many integer programming and combinatorial optimization problems remain challenging from a computational standpoint: they are NP-complete or worse and it is widely believed that no general and efficient algorithm exists for solving them. Solving them thus requires considerable time and expertise in both the application domain (modeling) and algorithm design (solving). In addition, the resulting algorithmic solutions often involve substantial development effort, since the distance between the problem model and the computer algorithm may be large.

This book describes OPL, a modeling language for combinatorial optimization that aims at simplifying the solving of these optimization problems. OPL was motivated by modeling languages such as AMPL and GAMS that provide computer equivalents to traditional algebraic notation. It provides similar support for modeling linear and integer programs and provides access to state-of-the-art linear programming algorithms. But OPL adds several new dimensions to modeling languages beyond the traditional support for linear and integer programming.

1.1 Background

To understand the novel aspects of OPL, it is useful to understand the strengths and weaknesses of the two technologies from which it is derived.

1.1.1 Modeling Languages

Modeling languages such as AMPL and GAMS were motivated by the desire to simplify the solving of mathematical programming problems. The fundamental insight underlying traditional modeling languages is the recognition that many mathematical programming problems can be expressed in a computer language whose syntax is close to the standard presentation of these problems in textbooks and scientific papers. These languages typically provide a number of data types such as arrays and sets, as well as computer-language equivalents to traditional algebraic notations. For instance, in AMPL, an expression such as

$$\sum_{i=1}^{n} a_i x_i$$

can be written as

```
sum {i in 1..n} a[i] * x[i]
```

In addition, some of these languages provide a clean separation between the model and the instance data. Finally, they are sometimes extended by a command language that makes it possible to solve sequences of related models and to make modifications to the models and solve the modified models.

Traditional modeling languages have many benefits that make them appealing for stating and solving mathematical programming problems. Perhaps their most significant contribution is to provide a language that directly supports the natural statement of these problems. This language abstracts away the implementation details of the underlying solver and users are then relieved of mundane, low-level, considerations and can focus on the modeling of their applications. Also important is the clear separation between the model and the instance data, which ensures that the same model can be applied to many instances without inducing additional work. Note that, in these languages, the solver is a black box that can only accessed through a set of well-defined parameters.[1]

Traditional modeling languages are particularly strong in mathematical programming applications, e.g., linear and integer programming. This is not surprising since this is the area from where they emerged. In addition, these problems are naturally expressed using traditional algebraic notations and effective solvers are available to solve the resulting models. However, a number of combinatorial optimization applications, such as job-shop scheduling and a variety of resource allocation problems, are outside the scope of these languages for a variety of reasons. On the one hand, these problems are rarely expressed naturally using algebraic constraints. On the other, it is often important in these applications to guide the solver towards solutions, or good solutions, by specifying an appropriate search procedure.

1.1.2 Mathematical Programming

Linear programming [6] is a important tool for combinatorial search problems, not only because it solves efficiently a large class of important problems, but also because it is the basic block of some fundamental techniques in this area. A linear program consists of minimizing a linear objective function subject to a set of linear constraints over real variables constrained to be nonnegative or, in symbols,

$$
\begin{array}{ll}
\text{minimize} & \sum_{j=1}^{n} c_j x_j \\
\text{subject to} & \\
& \sum_{j=1}^{n} a_{ij} x_j = b_i \quad (1 \leq i \leq m) \\
& x_j \geq 0 \quad (1 \leq j \leq n).
\end{array}
$$

Note first that considering only equations, nonnegative variables, and minimization is not restrictive. An inequality $t \geq 0$ can be recast as an equation $t - s = 0$ by adding a new variable, an arbitrary

---

[1] Note that one of the motivations for the development of AMPL and GAMS was to support the solution of nonlinear programming problems, which often require different kinds of solvers for different kinds of problems. Hence, the implementation of these mathematical programming modeling languages maintains a strong separation between the software processing the modeling language and the software implementing the solver. This makes it possible to use different solvers for different problems.

variable can be expressed as the difference of two nonnegative variables, and maximization can be expressed by negating the objective function. In addition, decision problems (i.e., finding if a set of constraints is satisfiable) can be recasted by adding a variable per constraint and minimizing their sum. The problem is satisfiable if and only if the optimum is zero. Note also that linear programs can be solved in polynomial time and robust solvers are now available that solve large scale linear programs.

The success of linear programming led many researchers to investigate some of its generalizations. Integer programming [11] is a natural extension of linear programming where variables are required to take integer values, i.e.,

$$
\begin{array}{ll}
\text{minimize} & \sum_{j=1}^{n} c_j x_j \\
\text{subject to} & \\
& \sum_{j=1}^{n} a_{ij} x_j = b_i \quad (1 \leq i \leq m) \\
& x_j \geq 0 \quad (1 \leq j \leq n) \\
& x_j \text{ integer} \quad (1 \leq j \leq n).
\end{array}
$$

Unfortunately, these integrality constraints make the problem NP-complete. Integer programming have been investigated extensively in the past decades and good solvers are now available for various classes of integer programs. However, many integer programs remain challenging from a computational standpoint.

Nonlinear programming is another generalization of linear programming that amounts to minimizing a nonlinear function subject to nonlinear constraints, i.e.,

$$
\begin{array}{ll}
\text{minimize} & g(x_1, \ldots, x_n) \\
\text{subject to} & f_1(x_1, \ldots, x_n) \geq 0 \\
& \ldots \\
& f_m(x_1, \ldots, x_n) \geq 0
\end{array}
$$

where $g, f_1, \ldots, f_n$ are real functions of n variables. Nonlinear programs are generally very challenging from a computational standpoint; local methods are often used to solve them, sacrificing optimality for speed of execution. Note also that integer programs can be recasted as nonlinear programs.

1.1.3 Constraint Programming

Constraint programming should not be viewed as an attempt to generalize linear programming, integer programming, or nonlinear programming. The term *programming* in constraint programming refers to its roots in the field of programming languages. As is well-known, there are various programming paradigms, e.g., procedural programming, object-oriented programming, functional programming, and logic programming and each of them has advantages and inconveniences for various classes of problems. Constraint programming is a recent entry in the field of programming languages that attempts to reduce the gap between the high-level description of an optimization problem and the computer algorithm implemented to solve it. Overviews of constraint programming, or subfields of constraint programming, can be found in [13, 26, 29, 34, 31], to name only a few; here we attempt merely to convey the main principles and benefits of this new technology.

What is Constraint Programming? The essence of constraint programming is a two-level architecture integrating a *constraint* and a *programming* component. The constraint component provides the basic operations of the architecture and consists of a system reasoning about fundamental properties of constraint systems such as satisfiability and entailment. The constraint component is often called the *constraint store*, by analogy to the memory store of traditional programming languages. The constraint store contains the constraints accumulated at some computation step and supports various queries and operations over these constraints. Operating around the constraint store is a programming-language component that specifies how to combine the basic operations, often in nondeterministic ways, since search is so fundamental in combinatorial optimization. The constraint and programming components can take many different forms depending upon the constraint system selected (e.g., linear constraints over reals) and the host programming languages (e.g., Prolog, C++).

The constraint systems featured in early constraint programming languages such as Prolog II [5], Prolog III [4], CHIP [8, 28], CLP($\Re$) [14], and BNR-Prolog [17] included constraint systems based on linear programming, consistency techniques, interval reasoning, and Boolean unification. Recent constraint languages, or new versions of the pioneering languages, now include many new algorithms tailored to certain application areas. Typical examples include the edge-finder algorithm and its generalization for the scheduling applications and flow algorithms that are fundamental in a variety of resource-allocation applications.

Since constraint programming originated from constraint logic programming, early programming components were based on the nondeterministic, goal-directed, computational model of Prolog. Concurrent constraint programming [15, 26], which is the foundation of constraint languages such as Oz [27] and cc(FD) [35], introduced a model in which the programming component is a set of agents communicating by adding constraints to, and querying, the constraint store. Concurrent constraint programming introduced constraint entailment as a basic operation and a constraint-driven computational model. Both of these features are now standard constraint programming technologies. Another important step in the development of modern constraint programming was the embedding of constraints in more traditional languages such as C and C++, as demonstrated, for instance, by Ilog Solver [20, 21] and 2LP [16]. Here the programming component is a traditional programming language extended to support nondeterminism and to support constraint systems as a basic data type.

Benefits of Constraint Programming Constraint programming is an appealing technology in a variety of combinatorial search problems, since it can reduce the development time of these applications by orders of magnitude. The gain in productivity comes mainly from the high-level abstractions provided by constraint programming to support two fundamental activities of combinatorial optimization algorithms: constraint reasoning and search. For constraint reasoning, constraint programming languages embed sophisticated constraint-solving algorithms that are accessed simply by specifying a set of constraints to satisfy. For search, constraint programming languages provide nondeterministic constructs that relieve programmers of many mundane implementation aspects of tree-search procedures. As a consequence, solving a combinatorial optimization problem using a constraint programming language generally has two steps: generating the set of constraints to be satisfied (the constraint part) and describing how to search for solutions (the search part).

The gain in productivity in constraint programming also comes from its declarative nature; a feature of course fundamental to modeling languages as well. Constraints specify properties of the solutions and can be thought of intuitively as restrictions on a space of possibilities. However, they do not specify a computational procedure for finding these solutions: i.e., constraints describe *what* the solutions are without specifying *how* to find them. This declarative nature of constraints has many benefits: the order in which constraints are imposed has no importance and additional constraints, that capture new properties of the solutions, can be added without worrying about the interaction with existing constraints and the search procedure. As a consequence, users can focus on the modeling aspects of the problem rather than on low-level programming details.

Finally, the separation of the constraint and search parts also has some obvious software engineering benefits. It is possible to modify these components independently, e.g., one can change the search strategy without having to be concerned about the constraint-solving part.

Limitations of Constraint Programming For many applications, however, the distance between the model and the constraint program is still large, mainly due to peculiarities of the host programming languages and lack of support for modeling. Constraint programming emerged from research on programming languages and there is a reluctance to sacrifice the general-purpose nature of these languages and a tendency to adopt a minimalist approach to extensions. However, the obvious trend at this point is to remedy this situation and to propose higher-level modeling tools.

Misconceptions about Constraint Programming Before concluding this section, it is important to dispell two common misconceptions about constraint programming. The first misconception is that constraint programming is an extension of mathematical programming. In fact, mathematical programming and constraint programming really address orthogonal issues that arise in solving combinatorial optimization problems. Mathematical programming focuses on identifying classes of problems, studying their properties, and proposing algorithms for solving them. In constrast, constraint programming is concerned with proposing software architectures (i.e., ways of organizing

computer programs) to simplify the implementation of combinatorial optimization algorithms. The
goal here is to shorten the development time of combinatorial optimization algorithms by reducing
the distance between a high-level design and an actual implementation.

A second common misconception is that constraint programming does not use bounding techniques in optimization problems. In fact, constraint programming uses a variety of bounding techniques (e.g., linear programming relaxations or preemptive scheduling) and can combine several bounding techniques naturally. This misconception probably orginates from the implicit nature of bounding procedures in constraint programming where bounding procedures are implemented as constraints that update the domains (e.g., the set of possible values) of the variables. For instance, to define a new bounding procedure in constraint programming, a new constraint of the form $c(x_1, \ldots, x_n, f)$ can be defined where $x_1, \ldots, x_n$ are the problem variables and f is the objective function. For a minimization problem, this constraint may add simpler constraints of the form $f \geq l$ at each node of the search tree (and, possibly, other constraints on the variables $x_1, \ldots, x_n$).

1.2 OPL

OPL is an attempt to combine the strengths of mathematical programming modeling languages and constraint programming. It aims both at increasing the applicability of modeling languages by incorporating techniques from constraint programming and at improving the expressive power of traditional constraint programming tools by borrowing ideas from modeling languages.

1.2.1 OPL as a Modeling Language

OPL shares many features with modeling languages such as AMPL. It supports traditional algebraic notations : a constraint such as

$$\sum_{j=1}^{n} d_{ij} = s$$

is written in OPL as

```
sum(j in 1..n) d[i,j] = s;
```

Similarly, the set of constraints

$$\sum_{j=1}^{n} d_{ij} = s \quad (1 \leq i \leq n)$$

is written in OPL as

```
forall(i in 1..n) sum(j in 1..n) d[i,j] = s;
```

1. Introduction

These features and others ensure that OPL preserves the traditional strengths of modeling languages in linear and integer programming.

However, OPL goes beyond the traditional algebraic support of modeling languages by introducing logical combinations of constraints, higher-order constraints, and enumerated types for variables, and by letting indices of arrays contain variables.[2] For instance, the OPL statement

```
enum Country {Belgium,Denmark,France,Germany,Netherlands,Luxembourg};
enum Colors {red,blue,yellow,gray};
var Colors color[Country];
solve {
   color[France] <> color[Belgium];
   color[France] <> color[Luxembourg];
   color[France] <> color[Germany];
   color[Luxembourg] <> color[Germany];
   color[Luxembourg] <> color[Belgium];
   color[Belgium] <> color[Netherlands];
   color[Belgium] <> color[Germanay];
   color[Germany] <> color[Netherlands];
   color[Germany] <> color[Denmark]
};
```

colors a map of several European countries so that no two adjacent countries have the same color. The statement specifies the countries and the colors, declares an array of variables (one per country) that may take four colors, and states the constraints. It is difficult to imagine a much more concise statement of this problem. Note, however, that the statement involves variables ranging over enumerated values and "not equal" constraints, two features usually not supported in mathematical programming modeling languages. OPL also supports arbitrary logical combinations such as

```
rankMen[m,o] < rankMen[m,wife[m]] => rankWomen[o,husband[o]] < rankWomen[o,m];
```

which illustrates the implication of two constraints. The example also shows the indexing of an array with variables or expressions involving variables, a very expressive modeling tool. In the above example, wife[m] and husband[o] are variables. Higher-order constraints are also a fundamental tool for modeling many applications. For instance, the expression

```
sum(j in Series) (s[j] = i)
```

---

[2] Some of these features, all standard in constraint programming, have in fact recently been proposed as possible extensions of AMPL.

counts the number of variables `s[j]` equal to i. Finally, OPL offers a rich set of modeling concepts for scheduling applications by including concepts such as activities and resources, also present in some constraint programming languages.

By providing these novel modeling tools, OPL adds a new dimension to both modeling and constraint programming languages. It should enlarge the applicability of modeling languages and make constraint programming techniques more accessible.

1.2.2 OPL as a Constraint Programming Language

Perhaps the most fundamental departure from mathematical programming modeling languages in OPL is its support for the programming component of the constraint programming architecture. OPL lets users specify search procedures using a variety of nondeterministic and constraint-driven constructs inspired by the development of constraint programming. As mentioned, this ability to specify how to explore the search space is fundamental to obtaining a reasonable efficiency in many problems and OPL provides high-level abstractions to support concise and effective specifications. Of course, OPL clearly separates the constraint and programming components, preserving and amplifying the spirit of constraint programming. A typical programming component in OPL is an instruction specifying what variables must be given values and in what order. For instance, the instruction

```
forall(s in Stores ordered by decreasing regretdmin(cost[s]))
    tryall(w in Warehouses ordered by increasing transportation[s,w])
      supplier[s] = w;
```

specifies a maximal regret heuristics in a warehouse location problem that is studied in detail in Chapter 10. OPL also supports constraint-driven constructs inspired by concurrent constraint programming.

By supporting both the constraint and the programming components of the constraint programming architecture, OPL adds a new dimension to modeling languages. No modeling language we are aware of supports the programming component.[3]

1.3 Contents

This book is a comprehensive presentation of OPL and is organized in two parts. Part I describes the language itself; each chapter in this part reviews some aspect of OPL in detail. These technical

---

[3] Note that the programming component discussed here is fundamentally differently from the command language of AMPL. The command language in AMPL is used to allow users to solve a sequence of related problems and to design algorithms that solve related models (by calling external solvers to solve single instances of a model). The goal of the programming component in OPL is to specify the search strategy for a model, i.e., how to explore the search space of possible solutions. These two extensions are orthogonal and complementary.

chapters are preceded by a short tour of OPL to introduce the main concepts and the computation model. Part II applies OPL in three application areas: linear and integer programming, constraint programming, and scheduling.

1.4 Model Conventions and Disclaimers

OPL statements and excerpts are printed in `tt` font and displayed in floating and numbered statements or enclosed between horizontal brackets, as in

```
var int nbRabbits in 0..20;
var int nbPheasants in 0..20;
solve {
   20 = nbRabbits + nbPheasants;
   56 = 4*nbRabbits + 2*nbPheasants;
};
```

The examples used in this book are intended to illustrate the functionalities of OPL. They should *not* be viewed as recommended or ultimate solutions for these problems. Indeed, better solutions can often be designed in some of the applications by exploiting more properties. These better solutions are, however, outside the scope of this book. Note also that the results displayed in this book are given only for the convenience of the readers and are not guaranteed to match exactly those of a specific implementation of OPL STUDIO.

I THE LANGUAGE

2 A Short Tour of OPL

This chapter, a brief tour of OPL, aims not to cover all the features of OPL but rather to give readers a preliminary understanding of the language and its novel aspects. The chapter starts by presenting how OPL supports linear programming and its extensions, an area that is well covered in traditional modeling languages. The chapter then turns to the support provided by OPL for constraint programming and scheduling, two areas new in modeling languages.

2.1 Linear and Integer Programming

An optimization problem is typically specified by an objective function and a set of constraints over some variables. A solution to the problem is an assignment of values to the variables that satisfies the constraints and optimizes the value of the objective function. The purpose of an OPL statement is thus to express these two components for the application at hand.

Consider a Belgian company Volsay, which specializes in producing ammoniac gas (NH_3) and ammonium chloride (NH_4Cl). Volsay has at its disposal 50 units of nitrogen (N), 180 units of hydrogen (H), and 40 units of chlorine (Cl). The company makes a profit of 40 Belgian francs for each sale of an ammoniac gas unit and 50 Belgian francs for each sale of an ammonium chloride unit. Volsay would like a production plan maximizing its profits given its available stocks. The OPL statement

```
var float+ gas;
var float+ chloride;

maximize
   40 * gas + 50 * chloride
subject to {
   gas + chloride <= 50;
   3 * gas + 4 * chloride <= 180;
   chloride <= 40;
};
```

formalizes this problem. It declares two real variables **gas** and **chloride**, representing the production of ammoniac gas and ammonium chloride. These variables are of type **float+**, which means that they are required to be nonnegative. The objective function

```
maximize 40 * gas + 50 * chloride
```

states that the profit must be maximized. The constraints ensure that the production plan does not exceed the available stocks of nitrogen, hydrogen, and chlorine, respectively. The constraint **gas +**

chloride <= 50 represents the capacity constraint for nitrogen, since each unit of ammoniac gas and of ammonium chloride uses one unit of nitrogen. The next two constraints, for hydrogen and chlorine respectively, are similar in nature.

As mentioned previously, a solution to an optimization problem is typically an assignment of values to the variables that satisfies the constraints and optimizes the objective function. OPL returns the optimal solution

```
Optimal Solution with Objective Value 2300.0000
  gas = 20.0000
  chloride = 30.0000
```

for the Volsay production-planning problem. This OPL statement is a linear programming model. As mentioned, linear programming is the class of problems that can be expressed as the optimization of a linear objective function subject to a set of linear constraints (i.e., linear equations and inequalities) over real numbers. Linear programming models can be solved for large numbers of variables and constraints and are, from a computational standpoint, the simplest applications considered in this book.

2.1.1 Arrays

The above statement is very specific to the application at hand. In general, it is desirable to write generic models that can be extended, modified easily, and applied in different contexts. The next two sections describe a number of OPL concepts to simplify the process of creating such models. A first step towards more genericity is the use of arrays, which makes it easier, for instance, to accommodate new products in the future. The Volsay production-planning model can be rewritten using arrays as:

```
enum Products {gas, chloride};
var float+ production[Products];
maximize
   40 * production[gas] + 50 * production[chloride]
subject to {
   production[gas] + production[chloride] <= 50;
   3 * production[gas] + 4 * production[chloride] <= 180;
   production[chloride] <= 40;
};
```

This new statement illustrates several features of the language. First, the instruction

```
enum Products {gas, chloride};
```

declares an enumerated set `Products` that represents the set of products of the company. The declaration

```
var float+ production[Products];
```

declares an array of two variables, `production[gas]` and `production[chloride]`, to represent the optimal production of ammoniac gas and ammonium chloride. These variables are used in the rest of the statement, which remains essentially the same as before. As will become clear subsequently, one of the novel features of OPL is the genericity of its arrays: OPL arrays can have an arbitrary number of dimensions and their index sets can be arbitrary finite sets, possibly involving complex data structures.

2.1.2 Data Declarations

A second fundamental step towards more genericity in the model amounts to representing the problem data explicitly. In addition to the products, the problem data obviously consists of the components (i.e., nitrogen, hydrogen, and chlorine), the demand of each product for each component, the profit of each product, and the stock available for each component. The following excerpt from an OPL statement

```
enum Products {gas, chloride};
enum Components {nitrogen, hydrogen, chlorine};

float+ demand[Products,Components] = [[1, 3, 0], [1, 4, 1]];
float+ profit[Products] = [40, 50];
float+ stock[Components] = [50, 180, 40];
```

declares and initializes these data. `Components` is an enumerated set that defines the chemical components necessary for the products, `demand` is a two-dimensional array whose element `demand[p,c]` represents the demand of product `p` for component `c`, and `profit` and `stock` are two arrays representing the profit of each product and the stock available for each component. The rest of the statement can be obtained easily by replacing the numbers by the relevant data items. For instance, the objective function is simply written as

```
maximize profit[gas]* production[gas] + profit[chloride] * production[chloride]
```

2.1.3 Aggregate Operators and Quantifiers

It should be clear, however, that the statement above contains much redundancy. All constraints, and all arithmetic terms in these constraints and in the objective function, are similar: they differ

```
enum Products {gas, chloride};
enum Components {nitrogen, hydrogen, chlorine};

float+ demand[Products,Components] = [[1, 3, 0], [1, 4, 1]];
float+ profit[Products] = [40, 50];
float+ stock[Components] = [50, 180, 40];

var float+ production[Products];
maximize
   sum(p in Products) profit[p] * production[p]
subject to {
   forall(c in Components)
      sum(p in Products) demand[p,c] * production[p] <= stock[c]
};
```

Statement 2.1: A Simple Production Model (gas1.mod).

only in their indices. OPL has two features to factorize these commonalities, aggregate operators and quantifiers, which are used in the new model in Statement 2.1. The objective function

```
maximize sum(p in Products) profit[p] * production[p]
```

illustrates the use of the aggregate operator `sum` to take the summation of the individual profits. A variety of aggregate operators are available in OPL, including `sum`, `prod`, `min`, and `max`. The instruction

```
forall(c in Components)
   sum(p in Products) demand[p,c] * production[p] <= stock[c]
```

shows how the universal quantifier `forall` can be used to state closely related constraints. It generates one constraint for each chemical component, each constraint stating that the total demand for the component cannot exceed its available stock. OPL supports rich parameter specifications in aggregate operators and quantifiers, as will become clear in Chapter 4.

2.1.4 Isolating the Data

Another fundamental step in making models reusable is to separate the model and the instance data. OPL STUDIO supports this clean separation through the notions of *projects*. A project is the association

```
enum Products ...;
enum Components ...;

float+ demand[Products,Components] = ...;
float+ profit[Products] = ...;
float+ stock[Components] = ...;

var float+ production[Products];
maximize
   sum(p in Products) profit[p] * production[p]
subject to {
   forall(c in Components)
      sum(p in Products) demand[p,c] * production[p] <= stock[p]
};
```

Statement 2.2: The Production Model (gas.mod).

of a model (a file whose suffix is .mod) and a set of data files (files whose suffixes are .dat). The model declares the data but does not initialize them. The data files contain the initializations of each declared data item. Here we do not describe the details of OPL STUDIO, but generally describe applications by giving the model and the instance data separately. For instance, Statements 2.2 and 2.3 together describe a project for the Volsay production planning problem. The model part is essentially the same as before, except that it declares the data but does not initialize it. A declaration of the form

```
float+ profit[Products] = ...;
```

declares the array profit and specifies that its initialization is given in a data file. The data file simply associates an initialization with each non-initialized piece of data.

2.1.5 Data Initialization

OPL offers a variety of ways to initialize data. One particularly useful feature is the possibility of associating indices with values to avoid various kinds of errors. Statement 2.4 illustrates this feature on the instance data for the Volsay production model. The initialization

```
profit = #[gas:30 chloride:40]#;
```

```
Products = {gas, chloride};
Components = {nitrogen, hydrogen, chlorine};
demand = [[1, 3, 0], [1, 4, 1]];
profit = [40, 50];
stock = [50, 180, 40];
```

Statement 2.3: Instance Data for the Production Model (gas.dat).

```
Products = {gas chloride};
Components = {nitrogen hydrogen chlorine};

profit = #[gas:30 chloride:40]#;
stock = #[nitrogen:50 hydrogen:180 chlorine:40]#;
demand = #[
   gas:      #[nitrogen:1 hydrogen:3 chlorine:0]#
   chloride: #[nitrogen:1 hydrogen:4 chlorine:1]#
   ]#;
```

Statement 2.4: Instance Data for the Production Model (gasn.dat).

describes the initialization of array profit by associating the value 30 with index gas and the value 40 with index chloride. (Of course, the order of the pairs has no importance in these initializations.) When using index:value pairs, the delimiters #[and]# must be used instead of [and]. Note also that, in data files, the items can be initialized in any order and the commas can be omitted freely.

2.1.6 Records

OPL offers a variety of data structures in addition to arrays and enumerated sets. Records, a fundamental tool for structuring the application data, offer an alternative to the traditional approach of representing data in parallel arrays. To see the use of records in OPL, consider the following production-planning model. To meet the demands of its customers, a company manufactures its products in its own factories (*inside* production) or buys them from other companies (*outside* pro-

2. A Short Tour of OPL

```
enum Products ...;
enum Resources ...;

float+ consumption[Products,Resources] = ...;
float+ capacity[Resources] = ...;
float+ demand[Products] = ...;
float+ insideCost[Products] = ...;
float+ outsideCost[Products] = ...;

var float+ inside[Products];
var float+ outside[Products];

minimize
   sum(p in Products) (insideCost[p]*inside[p] + outsideCost[p]*outside[p])
subject to {
   forall(r in Resources)
      sum(p in Products) consumption[p,r] * inside[p] <= capacity[r];
   forall(p in Products)
      inside[p] + outside[p] >= demand[p];
};
```

Statement 2.5: A Production-Planning Problem (production.mod).

duction). Inside production is subject to some resource constraints: each product consumes a certain amount of each resource. In contrast, outside production is theoretically unlimited. The problem is to determine how much of each product should be produced inside and outside the company while minimizing the overall production cost, meeting the demand, and satisfying the resource constraints.

Statement 2.5 depicts an OPL model for this problem that uses only the concepts introduced so far, and Statement 2.6 presents the data for a specific instance. An instance of the problem must specify the products, the resources, the capacity of the resources, the demand for each product, the consumption of resources by the different products, and the inside and outside costs of each product. These various data items are specified in the standard way in Statement 2.5. The model contains two arrays of variables: inside and outside. Element inside[p] (resp. outside[p]) represents the inside (resp. outside) production of product p. The objective function specifies that

```
Products = {kluski capellini fettucine};
Resources = {flour eggs};
consumption = [
      [0.5 0.2]
      [0.4 0.4]
      [0.3 0.6]
   ];
capacity = [20, 40];
demand = [100, 200, 300];
insideCost = [0.6, 0.8, 0.3];
outsideCost = [0.8, 0.9, 0.4];
```

Statement 2.6: Data for the Production-Planning Problem (production.dat).

the production cost must be minimized. The production cost is simply the sum of the individual production costs, which are obtained by multiplying the inside and outside productions of the given product by their respective costs. Finally, the model has two types of constraints. The first set of constraints expresses the capacity constraints, the second set states the demand constraints. The model is once again a linear programming problem and its output is as follows:

```
Optimal Solution with Objective Value 372.0000
   inside[kluski] = 40.0000
   inside[capellini] = 0.0000
   inside[fettucine] = 0.0000

   outside[kluski] = 60.0000
   outside[capellini] = 200.0000
   outside[fettucine] = 300.0000
```

Although the model is simple, it is inconvenient in separating the data associated with each product in different arrays: for instance, array **demand** stores the demand for the products, while array **insideCost** stores their inside costs. This technique, sometimes called *parallel arrays*, may be error-prone and less readable for more complicated models. Records provide a simple way to cluster related data and impose more structure on a model. This is illustrated in Statements 2.7 and 2.8, which exhibit an alternative model for the production- planning problem.

```
enum Products ...;
enum Resources ...;

struct ProductData {
   float+ demand;
   float+ insideCost;
   float+ outsideCost;
   float+ consumption[Resources];
};

ProductData product[Products] = ...;
float+ capacity[Resources] = ...;

var float+ inside[Products];
var float+ outside[Products];

minimize
   sum(p in Products)
      (product[p].insideCost*inside[p] + product[p].outsideCost*outside[p])
subject to {
   forall(r in Resources)
      sum(p in Products) product[p].consumption[r] * inside[p] <= capacity[r];
   forall(p in Products)
      inside[p] + outside[p] >= product[p].demand;
};
```

Statement 2.7: The Production-Planning Problem Revisited (`product.mod`).

```
Products = {kluski capellini fettucine};
Resources = {flour eggs};
product =
   #[
      kluski    : <100, 0.6, 0.8, [0.5, 0.2]>
      capellini : <200, 0.8, 0.9, [0.4, 0.4]>
      fettucine : <300, 0.3, 0.4, [0.3, 0.6]>
   ]#;
capacity = [20, 40];
```

Statement 2.8: Data for the Revised Production-Planning Problem (`product.dat`).

The instruction

```
struct ProductData {
    float+ demand;
    float+ insideCost;
    float+ outsideCost;
    float+ consumption[Resources];
};
```

declares a record type having four fields. The first three fields, of type float, are used to represent the demand and costs of a product; the last field is an array representing the resource consumptions of the product. These fields are intended to hold all the data related to a given product. The instruction

```
ProductData product[Products] = ...;
```

declares an array of these records, one for each product. The initialization

```
product =
   #[
      kluski    : <100 0.6 0.8 [0.5 0.2]>
      capellini : <200 0.8 0.9 [0.4 0.4]>
      fettucine : <300 0.3 0.4 [0.3 0.6]>
   ]#;
```

2. A Short Tour of OPL

from Statement 2.8 specifies these various data items: records are initialized by giving values for each of their fields. It is of course possible to use a named initialization for the record, as shown in Statement 2.9, in which case the initialization is enclosed with #< and >#. Record fields can be obtained by postfixing the record with a dot and the field name. For instance, in the objective function

```
minimize
   sum(p in Products)
      (product[p].insideCost * inside[p] + product[p].outsideCost * outside[p])
```

The expression `product[p].insideCost` represents the field `insideCost` of the record `product[p]`. Similarly, in the constraint

```
forall(r in Resources)
   sum(p in Products) product[p].consumption[r] * inside[p] <= capacity[r];
```

The expression `product[p].consumption` represents the field `consumption` of record `product[p]`. This field is an array that can be subscripted in the traditional way.

2.1.7 Displaying Results

The statements presented so far did not specify elements of the solution which should be displayed. OPL, and OPL STUDIO in particular, offer a variety of ways to display and visualize the results of an application; Chapter 8 gives more detail on the display instructions. An interesting feature of OPL is the ability to display tuples of expressions. For instance, it is possible to enhance Statement 2.7 with an instruction

```
display(p in Products) <inside[p],outside[p]>;
```

to produce the output

```
Optimal Solution with Objective Value 372.0
  <inside[kluski],outside[kluski]> = <40.0,60.0>
  <inside[capellini],outside[capellini]> = <0.0,200.0>
  <inside[fettucine],outside[fettucine]> = <0.0,300.0>
```

This display makes it possible to visualize the inside and outside productions of a product simultaneously. OPL also makes it possible to visualize useful information about the model solution. For instance, the instruction

```
display(p in Products) <inside[p],inside[p].rc)>;
```

```
Products = {kluski capellini fettucine};
Resources = {flour eggs};
product =
   #[
      kluski:
         #< demand:100
            insideCost:0.6
            outsideCost:0.8
            consumption:[0.5 0.2]
         >#
      capellini:
         #< demand:200
            insideCost:0.8
            outsideCost:0.9
            consumption:[0.4 0.4]
         >#
      fettucine:
         #< demand:300
            insideCost:0.3
            outsideCost:0.4
            consumption:[0.3 0.6]
         >#
   ]#;
capacity = [20, 40];
```

Statement 2.9: Named Data for the Revised Production-Planning Problem (productn.dat).

displays both the inside production of a product and its reduced cost:

```
Optimal Solution with Objective Value 372.0
  <inside[kluski],reducedCost(inside[kluski])> = <40.0,0.0>
  <inside[capellini],reducedCost(inside[capellini])> = <0.0,0.06>
  <inside[fettucine],reducedCost(inside[fettucine])> = <0.0,0.02>
```

2.1.8 Integer Programming

The statements presented so far are all linear programming models. As mentioned previously, linear programs with very large numbers of variables and constraints can be solved efficiently. Unfortunately, this is no longer true when the variables are required to take integer values. *Integer programming*, the class of problems that can be expressed as the optimization of a linear function subject to a set of linear constraints over integer variables, is in fact NP-hard [10]. More important, perhaps, is the fact that the integer programs that can be solved in reasonable time are much smaller in size than their linear programming counterparts. There are exceptions, of course, and this book describes several important classes of integer programs which can be solved efficiently, but users of OPL should be warned that discrete problems are in general much harder to solve than linear programs.

A typical example of integer programs is the knapsack problem, which can be intuitively understood as follows. We have a knapsack with a fixed capacity (an integer) and a number of items. Each item has an associated weight (an integer) and an associated value (another integer). The problem consists of filling the knapsack without exceeding its capacity, while maximizing the overall value of its contents. A multi-knapsack problem is similar to the knapsack problem, except that there are multiple features for the object (e.g., weight and volume) and multiple capacity constraints. Statement 2.10 depicts a model for the multi-knapsack problem, while Statement 2.11 describes an instance of the problem.

This model has several novel features. It represents items and resources not by enumerated sets but rather by integers. In other words, the items (resp. the resources) are represented by successive integers starting at 1. The instructions

```
int nbItems = ...;
int nbResources = ...;
range Items 1..nbItems;
range Resources 1..nbResources;
```

declare the number of items and the number of resources, as well as two ranges, Items and Resources, to represent the set of items and the set of resources. The next three instructions

```
int nbItems = ...;
int nbResources = ...;
range Items 1..nbItems;
range Resources 1..nbResources;
int capacity[Resources] = ...;
int value[Items] = ...;
int use[Resources,Items] = ...;
int maxValue = max(r in Resources) capacity[r];

var int take[Items] in 0..maxValue;
maximize
   sum(i in Items) value[i] * take[i]
subject to
   forall(r in Resources)
      sum(i in Items) use[r,i] * take[i] <= capacity[r];
```

Statement 2.10: A Multi-Knapsack Model (knapsack.mod).

```
nbResources = 7;
nbItems = 12;
capacity= [18209 7692 1333 924 26638 61188 13360];
value= [96 76 56 11 86 10 66 86 83 12 9 81];
use = [
      [19 1 10 1 1 14 152 11 1 1 1 1]
      [0 4 53 0 0 80 0 4 5 0 0 0]
      [4 660 3 0 30 0 3 0 4 90 0 0]
      [7 0 18 6 770 330 7 0 0 6 0 0]
      [0 20 0 4 52 3 0 0 0 5 4 0]
      [0 0 40 70 4 63 0 0 60 0 4 0]
      [0 32 0 0 0 5 0 3 0 660 0 9]];
```

Statement 2.11: Data for The Multi-Knapsack Problem (knapsack.dat).

```
int capacity[Resources] = ...;
int value[Items] = ...;
int use[Resources,Items] = ...;
```

are similar to the data declarations presented earlier in this chapter. The array `capacity` represents the capacity of the resources, the array `value` the value of each item, and `use[r,i]` the use of resource `r` by item `i`. The next instruction

```
int maxValue = max(r in Resources) capacity[r];
```

is more interesting. It declares an integer `maxValue` whose value is given by an expression. We see later in Chapter 4 that OPL has many features for computing and preprocessing data, since this is fundamental in simplifying and improving the efficiency of many models. The instruction

```
var int take[Items] in 0..maxValue;
```

declares the problem variables: `take[i]` represents the number of times item `i` is selected in the solution. The variable is of type integer and is restricted to range in `0..maxValue`. The rest of the statement is rather standard and should raise no difficulty. Here is the solution to the instance specified in Statement 2.11:

```
Optimal Solution with Objective Value 261922
  take[1] = 0
  take[2] = 0
  take[3] = 0
  take[4] = 154
  take[5] = 0
  take[6] = 0
  take[7] = 0
  take[8] = 913
  take[9] = 333
  take[10] = 0
  take[11] = 6499
  take[12] = 1180
```

Although integer programs are, in general, substantially harder to solve than linear programs, they have also been the topic of intensive investigation. OPL recognizes when a statement is an integer programming model and applies an efficient integer programming algorithm.

2.1.9 Mixed Integer-Linear Programming

OPL can also solve models that include both integer and real variables, generally known as mixed integer-linear programs (MILP). OPL approaches them in essentially the same way as integer programs except, of course, that branching takes place only on integer variables. Consider the following application involving mixing some metals into an alloy. The metal may come from several sources: in pure form or from raw materials, scraps from previous mixes, or ingots. The alloy must contain a certain amount of the various metals, as expressed by a production constraint specifying lower and upper bounds for the quantity of each metal in the alloy. Each source also has a cost and the problem consists of blending the sources into the alloy while minimizing the cost and satisfying the production constraints. Similar problems arise in other domains, e.g., the oil, paint, and the food processing industries. Statements 2.12 and 2.13 depict a model for the problem, while Statement 2.14 depicts some instance data.

Model Data The model is described in terms of a number of constants specifying the various types of metals, raw materials, scrap, and ingots. In the instance, there are three metals, two raw materials, two kinds of scrap, and one kind of ingot. The model also defines ranges for each of the components. It then defines the cost of the various components in `costMetal`, `costRaw`, `costScrap`, `costIngo`. In the instance data, for example, the second raw material has a cost of 5. The data items `low` and `up` specify the production constraints and give lower and upper bounds on the quantity of each sort of metal in the alloy. For example, in the instance data, between 30% and 40% of the alloy must be the second metal. The next data items, `percRaw`, `percScrap`, and `percIngo`, specify the percentage of each metal in the sources. In our instance, for example, the second type of scrap contains 1% of the first metal, none of the second metal, and 70% of the third metal. Finally, the data `alloy` specifies the amount of alloy to be produced.

Model Variables The model variables specify how much of each source is used in the alloy: the array `p` specifies the quantities of pure metals, array `r` specifies the quantities of raw materials, array `s` specifies the quantities of scrap, array `i` specifies the number of ingots. All variables are of type float except number of ingots, which are integers. The problem is thus a mixed integer-linear program. The instruction

```
var float m[j in Metals] in low[j] * alloy .. up[j] * alloy;
```

is particularly interesting, since it shows how to specify the range of variables in a generic fashion. More precisely, the range of variables `m[j]` is given by the expression `low[j] * alloy .. up[j] * alloy`. Note also that the model uses the variables in array `m` as intermediary variables to represent the quantity of each metal produced.

```
int nbMetals = ...;
int nbRaw = ...;
int nbScrap = ...;
int nbIngo = ...;

range Metals 1..nbMetals;
range Raws 1..nbRaw;
range Scraps 1..nbScrap;
range Ingos 1..nbIngo;

float+ costMetal[Metals] = ...;
float+ costRaw[Raws] = ...;
float+ costScrap[Scraps] = ...;
float+ costIngo[Ingos] = ...;
float+ low[Metals] = ...;
float+ up[Metals] = ...;
float+ percRaw[Metals,Raws] = ...;
float+ percScrap[Metals,Scraps] = ...;
float+ percIngo[Metals,Ingos] = ...;

int+ alloy = ...;
```

Statement 2.12: A Blending Problem: Part I (blending.mod).

```
var float+ p[Metals];
var float+ r[Raws];
var float+ s[Scraps];
var int+ i[Ingos];
var float m[j in Metals] in low[j] * alloy .. up[j] * alloy;
minimize
   sum(j in Metals) costMetal[j] * p[j] + sum(j in Raws) costRaw[j] * r[j] +
   sum(j in Scraps) costScrap[j] * s[j] + sum(j in Ingos) costIngo[j] * i[j]
subject to {
   forall(j in Metals)
      m[j] = p[j] + sum(k in Raws) percRaw[j,k] * r[k] +
      sum(k in Scraps) percScrap[j,k] * s[k] + sum(k in Ingos) percIngo[j,k] * i[k]
   sum(j in Metals) m[j] = alloy;
};
```

Statement 2.13: A Blending Problem: Part II (blending.mod).

```
nbMetals = 3;
nbRaw = 2;
nbScrap = 2;
nbIngo = 1;
costMetal = [22 10 13];
costRaw = [6 5];
costScrap = [7 8];
costIngo = [9];
low = [0.05 0.30 0.60];
up  = [0.10 0.40 0.80];
percRaw = [[0.20 0.01] [0.05 0.00] [0.05 0.30]];
percScrap = [[0.00 0.01] [0.60 0.00] [0.40 0.70]];
percIngo = [[0.10] [0.45] [0.45]];
alloy = 71;
```

Statement 2.14: Instance Data for the Blending Problem (blending.dat).

2. A Short Tour of OPL

Model Constraints There are two types of constraints in this problem. The first

```
forall(j in Metals)
    m[j] = p[j] + sum(k in Raws) percRaw[j,k] * r[k] +
    sum(k in Scraps) percScrap[j,k] * s[k] + sum(k in Ingos) percIngo[j,k] * i[k]
```

makes sure that the right amounts of metal are produced. The amount m[j] of metal j must be equal to the amount of pure metal p[j] added to the quantity of metal j contained in the raw materials, the scrap, and the ingots. The correct amount of metals are computed using the percentage of metals contained in the sources. The last constraint

```
sum(j in Metals) m[j] = alloy;
```

makes sure that the various metals produced give the correct amount of alloy. The objective function in this model is rather simple. It consists of computing the price of each source from its unit price (e.g., costMetal) and the amount produced (e.g., p[j]). When given the model and the instance data, OPL returns the solution

```
Optimal Solution with Objective Value 653.6100
  p[1] = 0.0467
  p[2] = 0.0000
  p[3] = 0.0000

  r[1] = 0.0000
  r[2] = 0.0000

  s[1] = 17.4167
  s[2] = 30.3333

  i[1] = 32

  m[1] = 3.5500
  m[2] = 24.8500
  m[3] = 42.6000
```

2.2 Constraint Programming

The most novel aspect of OPL compared to other mathematical modeling languages lies in its support for constraint programming. Constraint programming, a recent entry in the programming languages arena, is in essence a two-level architecture consisting of

- *a constraint component:* a constraint-solving system reasoning about fundamental properties of constraints such as satisfiability and entailment;
- *a programming component:* a programming language specifying how to generate, combine, and process constraints, often in nondeterministic ways.

Constraint programming support in OPL is discussed in detail in Chapter 10. Here we give a brief introduction to its constraint programming features. Consider finding an eight digit number that is a square and remains a square when 1 is concatenated in front of its decimal notation. Here is a simple OPL solution to the problem:

```
var int n in 10000000..99999999;
var int x in 0..20000;
var int y in 0..20000;
solve {
   n = x*x;
   100000000+n = y*y;
};
```

The statement declares three variables: the variable n, which is the desired result, and two variables x and y, which are used to state the constraints. The range of variable n is chosen to represent the fact that n is a eight-digit number, while the ranges of x and y were chosen in order not to eliminate any solution. The `solve` instruction specifies the problem constraints in terms of the variables. Note that the two constraints are nonlinear and that there is no objective function.

When OPL encounters a statement that is not a linear, an integer, or a mixed integer linear program, its basic strategy is to use the problem constraints to reduce the domains (i.e., the set of possible values) of the variables. The constraints specify which combinations of these values are solutions of the problem and OPL uses various relaxations of these constraints to prune the search space, i.e., the domains of the variables. Sometimes constraint solving alone is sufficient to find a solution to the problem of interest. However, in general, the constraint-solving process does not locate a solution either because there are multiple solutions or because the constraint-solving algorithms are not strong enough, since they were designed as a tradeoff between computational complexity and pruning. For the above problem, constraint solving alone produces the following domain reductions:

```
n in [23765625..56250000]
x in [4875..7500]
y in [11125..12500]
```

by using reasoning on the range of the variables.

To find a solution to this problem, OPL must generate values for the variables. In particular, OPL assigns a value, say 23765625, to n and applies its constraint-solving algorithm with this new assumption. If constraint solving fails (i.e., if there is no solutions to this problem), OPL backtracks. Backtracking undoes all the changes to the domains of the variables to restore the same computational state as before the nondeterministic choice. OPL then continues its execution by trying another value for n. The process terminates when OPL finds a solution or determines that there are no solutions (if all possible assignments lead to failure). On the above example, OPL returns the solution

```
Solution [1]
  n = 23765625
  x = 4875
  y = 11125
```

It is also possible to ask OPL to produce all solutions or obtain other solutions incrementally. When asked to find all solutions, OPL returns the two solutions:

```
Solution [1]
  n = 23765625
  x = 4875
  y = 11125
Solution [2]
  n = 56250000
  x = 7500
  y = 12500
```

As can be seen, OPL has default strategies to search for solutions. However, for more advanced applications, it is often desirable to provide strategies tailored to the problem at hand. In the above example, the default strategy consists of generating a value for all variables, which is equivalent to the statement

```
var int n in 10000000..99999999;
var int x in 0..20000;
var int y in 0..20000;
solve {
   n = x*x;
   100000000+n = y*y
};
```

```
var int queens[1..8] in 1..8;
solve {
   forall(ordered i, j in 1..8) {
      queens[i] <> queens[j];
      queens[i] + i <> queens[j] + j;
      queens[i] - i <> queens[j] - j;
   }
};
```

Statement 2.15: The Eight-Queens Problem (queens8.mod).

```
search {
   generate(n);
   generate(x);
   generate(y);
};
```

The above statement includes a search procedure that tells OPL to generate values for n, x, and y. This is a fundamental novelty modeling languages: OPL makes it possible to state concisely not only the problem constraints but also how to search for solutions. An interesting example illustrating this novel functionality is described later in this chapter.

To understand constraint programming in more detail, it is useful to consider the n-queens problem, which consists of placing n queens on a chessboard of size $n \times n$ so that no two queens attack each other. Since no two queens can be placed on the same column, a simple model of this problem consists of associating a queen with each column and searching for an assignment of rows to the queens that no two queens are placed on the same row and the same diagonal. Statement 2.15 depicts an OPL statement for the eight-queens problems.

The statement has a number of interesting features. It first declares an array of eight variables, all of which take their value in the range 1..8. The `solve` instruction defines the problem constraints, i.e., that no two queens should attack each other. The basic idea here is to generate for all $1 \leq i < j \leq 8$ the constraints

```
queens[i] <> queens[j]
queens[i] + i <> queens[j] + j
queens[i] - i <> queens[j] - j
```

2. A Short Tour of OPL

```
int n << "number of queens:";
range Domain 1..n;

var Domain queens[Domain];
solve {
   forall(ordered i,j in Domain) {
      queens[i] <> queens[j];
      queens[i] + i <> queens[j] + j;
      queens[i] - i <> queens[j] - j
   }
};
```

Statement 2.16: The n-Queens Model (`queens.mod`).

where the symbol `<>` means "not-equal". OPL uses a single universal quantifier for parameters `i` and `j` and specifies that these parameters are ordered, i.e., `i < j`. Such a construct is useful in many applications to make statements shorter and more explicit. Note also that the statement is not an integer programming model, since integer programming only supports linear equations and inequalities (and not strict inequalities such as `<`, `>`, and `<>`). Statement 2.16 generalizes the problem to an arbitrary number of queens. It declares an integer that, when executed, prints the message *Number of queens:* and requests an integer from the user.

It is interesting at this point to study how OPL behaves on this problem. To illustrate the computational model, consider first the five-queens problem. Constraint solving does not reduce the domains of the variables initially. OPL thus generates a value, say 1, for one of its variables, say `queens[1]`. After this nondeterministic assignment, constraint solving removes inconsistent values from the domains of `queens[2]`,...,`queens[5]`, as depicted in Figure 2.1. The next step of the generation process tries the value 3 for `queens[2]`. OPL then removes inconsistent values from the domains of the remaining queens (see Figure 2.2). Since only one value remains for `queens[3]` and `queens[4]`, these values are immediately assigned by OPL to these variables and, after more constraint solving, OPL assigns the value 4 to `queens[5]`. A solution to the five-queens problem is thus found with two choices and without backtracking.

It is also interesting to consider the eight-queens problem after three choices. Figure 2.3 depicts an intermediate stage when some (but not all) constraint solving has taken place. Since only one value is left for `queens[6]`, OPL assigns this value and propagates the constraints further. This assigns the value 7 to `queens[8]` which, in turn, assigns the value 2 to `queens[7]` and the value 8

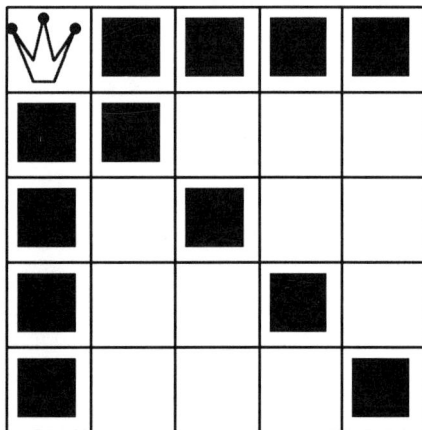

Figure 2.1: The Five-Queens Problem After One Choice.

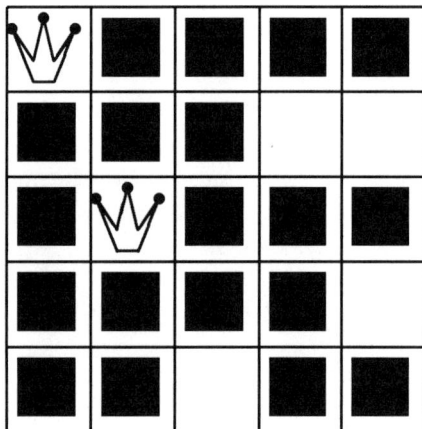

Figure 2.2: The Five-Queens Problem After Two Choices.

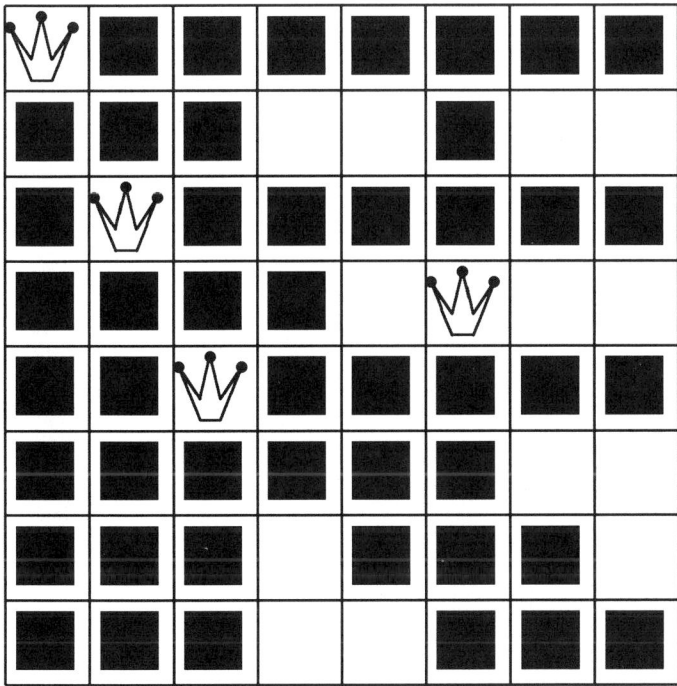

Figure 2.3: The Eight-Queens Problem After Three Choices (Intermediate).

Figure 2.4: The Eight-Queens Problem After Three Choices (Intermediate).

to queens[4] (see Figure 2.4). As a consequence, queens[5] has no value left in its domain and constraint solving fails. OPL then backtracks and tries another value for queens[3].

It is useful to mention at this point that OPL does not necessarily generate values for the queens in the order queens[1] to queens[n]. In fact, the default strategy of OPL (which is implemented in instruction **generate**) chooses first the variable with the fewest possible values. This heuristic, known as the *first-fail principle*, is in general effective in reducing the size of the search space by discovering failures early. To generate values for the queens in the order queens[1] to queens[n], it is sufficient to write the following search procedure:

```
search {
    forall(i in Domain)
        generate(queens[i]);
};
```

```
int n << "Number of Variables:   ";

range Range 0..n-1;
range Domain 0..n;

var Domain s[Range];
solve {
   forall(i in Range)
      s[i] = sum(j in Range) (s[j] = i);
};
```

Statement 2.17: The Magic-Series Model (`magic1.mod`).

2.2.1 Higher-Order Constraints

The magic-series problem has become a traditional constraint- programming benchmark and illustrates convincingly the use of higher-order constraints. The problem consists of finding a magic series, i.e., a sequence of numbers $S = (s_0, s_1, \ldots, s_n)$ such that s_i represents the number of occurrences of i in S. For instance, $(1, 2, 1, 0)$ is a magic series for $n = 3$, since there is one occurrence of 0, two occurrences of 1, one occurrence of 1 and zero occurrences of 3. Statement 2.17 depicts an OPL model for this problem.

The most interesting aspect of this statement is the constraint specification, which expresses the occurrence constraints using higher-order constraints. An expression

`sum(j in Range) (s[j] = i)`

is used to count the number of variables in `s[0]`, ..., `s[n-1]` equal to i. The novelty here is that the summation is over relations of the form `s[j] = i` and not over traditional expressions. The key insight to understanding this summation is to realize that the OPL implementation replaces each constraint appearing in an expression by a 0-1 variable (i.e., an integer variable ranging over 0..1) that is assigned the value 1 if the constraint is satisfied and 0 if it is violated. It is thus easy to see that the above statement exactly captures the problem constraints. In addition, constraints can be combined in OPL using the traditional logical connectives.

It is also interesting to see how OPL uses higher-order constraints during constraint solving. The key idea is to propagate information in both directions: from the constraint to its associated 0-1 variable and from the 0-1 variable to its associated constraint. In the first direction, OPL tries to determine whether the constraint is satisfied. The satisfiability of the constraint is determined

```
int n << "Number of Variables:   ";

range Range 0..n-1;
range Domain 0..n;

var Domain s[Range];
solve {
   forall(i in Range)
      s[i] = sum(j in Range) (s[j] = i);
   sum(i in Range) s[i] = n;
   sum(i in Range) s[i]*i = n;
};
```

Statement 2.18: The Magic-Series Problem With Redundant Constraints (magic2.mod).

locally on the basis of the domains of the variables. For instance, if x is a variable ranging over $5\ldots 10$ and y a variable ranging over $1\ldots 3$, OPL determines that x > y is satisfied, in which case its associated 0-1 variable is assigned the value 1. For the same domain, OPL would also determine that the constraint x < y is violated, in which case its associated 0-1 variable is assigned the value 0. However, if y ranges over $1\ldots 6$, OPL cannot determine whether x > y is satisfied or violated, since there exist assignments that satisfy the constraint and assignments that violate it. In the second direction, OPL uses the 0-1 variable to insert new constraints. For instance, if the 0-1 variable associated with x > y is assigned to 1 (by using the constraint in which the higher-order constraint appears), OPL inserts the constraint x > y. On the other hand, if the 0-1 variable associated with x > y is assigned to 0, OPL inserts the constraint x ≤ y.

Statement 2.18 depicts another OPL model for the magic-series problem, that contains two additional "redundant" (or surrogate) constraints. The constraints are redundant since they do not remove any solution compared to the previous statement. They are, however, interesting from a computational standpoint, since they express some fundamental properties of the problem that OPL cannot deduce automatically but can exploit by OPL to reduce the search space and obtain solutions faster. The first of these redundant constraints expresses that the sum of the elements in the series is n, which is obviously true since there are n variables. The second constraint expresses the same information but computes the sum differently by using products of the form s[i] * i (i.e., the value i times its number of occurrences).

Finally, we show a formulation of the magic-series problem using the global constraint distribute.

```
int n << "Number of Variables:   ";

range Range 0..n-1;
range Domain 0..n;

int value[i in Range] = i;

var Domain s[Range];
solve {
   distribute(s,value,s);
   sum(i in Range) s[i] = n;
   sum(i in Range) s[i]*i = n;
};
```

Statement 2.19: The Magic-Series Problem With the Global Constraint `distribute` (`magic3.mod`).

Global constraints, a fundamental tool for solving a variety of combinatorial optimization problems in OPL, enforce complex relationships among a number of variables. They are discussed in more detail in Chapter 5. Given a one-dimensional array `occurrence` of index set S, a one-dimensional array `value` with the same index set S, and a one-dimensional array `element`,

`distribute(occurrence,value,element)`

holds if `occurrence[i]` is the number of occurrences of `value[i]` in array `element`. The effect of `distribute` could be implemented by a conjunction of higher-order constraints but achieves the same pruning more efficiently. Statement 2.19 shows the magic-series problem with this global constraint. Besides the use of the higher-order constraint, Statement 2.19 also introduces a novel concept: the generic initialization of an array. The declaration

`int value[i in Range] = i;`

declares an array of n integers: `value[0]` to `value[n-1]`. These integers are assigned generically to the values 0 to $n-1$ by using a parameter `i` ranging over `Range`. As will become clear, generic initializations are also a fundamental tool in OPL for facilitating the model specification.

```
enum Country {Belgium,Denmark,France,Germany,Netherlands,Luxembourg};
enum Colors {blue,red,yellow,gray};
var Colors color[Country];
solve {
   color[France] <> color[Belgium];
   color[France] <> color[Luxembourg];
   color[France] <> color[Germany];
   color[Luxembourg] <> color[Germany];
   color[Luxembourg] <> color[Belgium];
   color[Belgium] <> color[Netherlands];
   color[Belgium] <> color[Germany];
   color[Germany] <> color[Netherlands];
   color[Germany] <> color[Denmark];
};
```

Statement 2.20: A Map-Coloring Problem (`map.mod`).

2.2.2 Variables over Enumerated Sets

In the models presented so far, variables ranged over integer or real values. In OPL, it is also possible to declare variables taking values in an enumerated set. This feature is interesting in avoiding the mapping from enumerated values into integers. Consider a map-coloring application entailing coloring a map of various European countries so that no two adjacent countries have the same color. Statement 2.20 gives an OPL model to solve this problem. The statement uses two enumerated sets to denote the countries and the colors (more colors are unnecessary since every map can be colored with at most four colors). The statement also declares an array of variables, each variable in the array ranging over the colors. The constraints specify that adjacent countries be assigned different colors. Here is the first solution returned by OPL:

```
Solution [1]
  color[Belgium] = blue;
  color[Denmark] = red;
  color[France] = red;
  color[Germany] = yellow;
  color[Netherlands] = red;
  color[Luxembourg] = gray;
```

	Richard	James	John	Hugh	Greg
Helen	1	2	4	3	5
Tracy	3	5	1	2	4
Linda	5	4	2	1	3
Sally	1	3	5	4	2
Wanda	4	2	3	5	1

Figure 2.5: Ranking of the Women in the Stable Marriage Problem.

2.2.3 Variables and Expressions as Indices

We now illustrate another feature of constraint programming: the use of variables, and of expressions involving variables, to index arrays. This feature is fundamental in obtaining concise and clear models for many combinatorial optimization problems. This section illustrates the gain in expressiveness on the stable marriage problem, and also shows that constraints can be combined with Boolean connectives. The similarity between the OPL statement and the natural language specification is striking.

The problem can be described as follows. Consider a group of women and a group of men who must marry and assume that each person has indicated a ranking for her/his possible spouses. The problem is to find a matching between the two groups such that the marriages are stable. The definition of stability is interesting: a marriage between m and w is stable provided that

- whenever m prefers another person o to w, o prefers her/his spouse to m.
- whenever w prefers another person o to m, o prefers her/his spouse to m.

Intuitively, m and w may be unhappy but they are bound to stay together. An instance of this problem is depicted in Figures 2.5 and 2.6, where a lower ranking indicates a stronger preference. For instance, the first row in Figure 2.5 says that Richard is Helen's first choice (rank 1) and, Greg is her last choice (rank 5). Statement 2.21 depicts an OPL model for the stable marriage problem and Statement 2.22 presents the data for this instance. The model data consists of two enumerated types for specifying the women and men and of rankings given by the women and by the men. The model is expressed in terms of two arrays of variables: `wife`, which specifies the men's wives, and `husband`, which specifies the women's husbands. Variables in array `wife` must take their values in enumerated set `Women`, while variables in array `husband` must take their values in set `Men`.

The constraints in this problem are of course the interesting and novel part. The first two sets of constraints guarantee that a solution is a set of marriages by ruling out the possibility that a man be married to a woman who is married to another man (and vice-versa). The model satisfies this requirement by stating the spouse of the spouse of a person is the person herself, i.e.,

	Helen	Tracy	Linda	Sally	Wanda
Richard	5	1	2	4	3
James	4	1	3	2	5
John	5	3	2	4	1
Hugh	1	5	4	3	2
Greg	4	3	2	1	5

Figure 2.6: Ranking of the Men in the Stable Marriage Problem.

```
enum Women ...;
enum Men ...;

int rankWomen[Women,Men] = ...
int rankMen[Men,Women] = ...

var Women wife[Men];
var Men husband[Women];
solve {
   forall(m in Men)
      husband[wife[m]] = m;

   forall(w in Women)
      wife[husband[w]] = w;

   forall(m in Men & o in Women)
      rankMen[m,o] < rankMen[m,wife[m]] => rankWomen[o,husband[o]] < rankWomen[o,m];

   forall(w in Women & o in Men)
      rankWomen[w,o] < rankWomen[w,husband[w]] => rankMen[o,wife[o]] < rankMen[o,w];
}
```

Statement 2.21: The Stable Marriage Problem (marriage.mod).

2. A Short Tour of OPL

```
Men = {Richard,James,John,Hugh,Greg};
Women = {Helen,Tracy,Linda,Sally,Wanda};
rankWomen =
   #[
      Helen :  #[Richard:1, James:2, John:4, Hugh:3, Greg:5]#,
      Tracy :  #[Richard:3, James:5, John:1, Hugh:2, Greg:4]#,
      Linda :  #[Richard:5, James:4, John:2, Hugh:1, Greg:3]#,
      Sally :  #[Richard:1, James:3, John:5, Hugh:4, Greg:2]#,
      Wanda :  #[Richard:4, James:2, John:3, Hugh:5 Greg:1]#
   ]#;

rankMen =
   #[
      Richard : #[Helen:5, Tracy:1 , Linda:2 , Sally:4 , Wanda:3]#,
      James   : #[Helen:4, Tracy:1 , Linda:3 , Sally:2 , Wanda:5]#,
      John    : #[Helen:5, Tracy:3 , Linda:2 , Sally:4 , Wanda:1]#,
      Hugh    : #[Helen:1, Tracy:5 , Linda:4 , Sally:3 , Wanda:2]#,
      Greg    : #[Helen:4, Tracy:3 , Linda:2 , Sally:1 , Wanda:5]#
   ]#;
```

Statement 2.22: Instance Data for the Stable Marriage Problem (`marriage.dat`).

```
forall(m in Men)
    husband[wife[m]] = m;
forall(w in Women)
    wife[husband[w]] = w;
```

It is important to recognize here that `husband` and `wife` are arrays of variables, so that an expression such as `husband[wife[m]]` indexes an array of variables with a variable. This feature is critical in this example to state the constraints in a simple way. The constraint-solving algorithm of OPL knows how to use these expressions to reduce the domains of the variables in the arrays `husband` and `wife` using bidirectional propagation.

The remaining constraints specify the stability requirement and are almost direct translations of the definition. The set of constraints

```
forall(m in Men & o in Women)
    rankMen[m,o] < rankMen[m,wife[m]] => rankWomen[o,husband[o]] < rankWomen[o,m];
```

states that, if a man `m` prefers a woman `o` to his wife (i.e., the rank of `o` is less than the rank of his wife), then `o` prefers her husband to `m`. The last set of constraints

```
forall(w in Women & o in Men)
    rankWomen[w,o] < rankWomen[w,husband[w]] => rankMen[o,wife[o]] < rankMen[o,w];
```

handles the symmetric case. The OPL model uses a logical implication together with the possibility of having variables index an array to make this statement simple and natural. Here is the first solution returned by OPL when given Statements 2.21 and 2.22:

```
Solution [1]
  wife[Richard] = Tracy
  wife[James] = Helen
  wife[John] = Wanda
  wife[Hugh] = Linda
  wife[Greg] = Sally

  husband[Helen] = James
  husband[Tracy] = Richard
  husband[Linda] = Hugh
  husband[Sally] = Greg
  husband[Wanda] = John
```

2.2.4 Search Procedures

As mentioned previously, the ability in constraint programming to specify the search procedures is often fundamental to obtaining reasonable efficiency or finding good solutions in hard combinatorial optimization problems. OPL support for search is described in detail in Chapter 7. This section illustrates how search procedures are implemented in OPL on a two-dimensional packing problem. The *perfect square* problem is to assemble a number of squares, all of different sizes, to produce a larger square, called the master square, in such a way that the squares do not overlap and leave no empty space. Placing 21 squares in a master square is of particular interest since 21 is the smallest number of squares that can produce another square.

Statement 2.23 depicts an OPL model for this problem, the data for which can be described concisely. Integer sizeSquare describes the size of the master square, while nbSquares defines the number of squares to pack. The range declarations define the ranges associated with these constants. The next declaration simply defines an array containing the sizes (i.e., the lengths of the sides) of the squares.

The variables in this problem are more interesting. The basic intuition is to associate with each square a point denoting the position of its bottom left corner. This identifies uniquely the space occupied by the square, since the size of the square is known. To implement this idea, the model uses two arrays of variables for the coordinates of a point, which denote of course the positions of the bottom left corners of the squares. There are three main groups of constraints in the model. The constraints

```
forall(s in Squares) {
    x[s] <= sizeM + 1 - size[s];
    y[s] <= sizeM + 1 - size[s];
}
```

simply state that the squares must fit in the master square by making sure there is enough room to the right of and above the bottom left corner. The constraints

```
forall(ordered i, j in Squares)
       x[i] + size[i] <= x[j]
    \/ x[j] + size[j] <= x[i]
    \/ y[i] + size[i] <= y[j]
    \/ y[j] + size[j] <= y[i];
```

express the fact that the squares may not overlap by using a disjunction of constraints. Two squares i and j do not overlap if square i is on the left

```
x[i] + size[i] <= x[j]
```

```
int sizeM = 112;
int NbSquares = 21;
range Squares 1..NbSquares,
range Positions 1..sizeM;
int size[Squares] = [50,42,37,35,33,29,27,25,24,19,18,17,16,15,11,9,8,7,6,4,2];
var Positions x[Squares],
var Positions y[Squares];

solve {
   forall(s in Squares) {
      x[s] <= sizeM + 1 - size[s];
      y[s] <= sizeM + 1 - size[s]
   };
   forall(ordered i, j in Squares)
         x[i] + size[i] <= x[j]
      \/ x[j] + size[j] <= x[i]
      \/ y[i] + size[i] <= y[j]
      \/ y[j] + size[j] <= y[i];
   forall(p in Positions) {
      sum(s in Squares) size[s] * (x[s] <= p & p <= x[s] + size[s] - 1) = sizeM;
      sum(s in Squares) size[s] * (y[s] <= p & p <= y[s] + size[s] - 1) = sizeM;
   }
};
search {
   forall(p in Positions)
      forall(s in Squares)
         try x[s] = p | x[s] <> p endtry;
   forall(p in Positions)
      forall(s in Squares)
         try y[s] = p | y[s] <> p endtry;
};
```

Statement 2.23: The Perfect Square Problem (`square.mod`).

on the right

```
x[j] + size[j] <= x[i]
```

below

```
y[i] + size[i] <= y[j]
```

or above square j

```
y[j] + size[j] <= y[i]
```

The last set of constraints

```
forall(p in Positions) {
    sum(s in Squares) size[s] * (x[s] <= p & p <= x[s] + size[s] - 1) = sizeM;
    sum(s in Squares) size[s] * (y[s] <= p & p <= y[s] + size[s] - 1) = sizeM
}
```

are redundant constraints that exploit the knowledge that there is no empty space in the master square. The basic intuition here consists of drawing a vertical (resp. horizontal) line and retrieving the squares intersecting the line. Since there is no empty space, the sizes of the intersecting squares must be equal to the size of the master square. The intersection for the vertical line is computed simply by testing if the line is between the x-coordinates of the left and right corners, i.e.,

```
(x[s] <= p & p <= x[s] + size[s] - 1)
```

These expressions should be viewed as 0-1 variables. Multiplying them by the size of the squares makes it possible to sum the sizes of all intersecting squares and it is then sufficient to state that the sum must be equal to the size of the master square.

The most novel aspect in this problem is, of course, the search procedure, which directly exploits the structure of the problem. The key idea is, once again, to use the fact that the master square has no empty space. As a consequence, the model considers each x- and y-position in the master square and chooses, in a nondeterministic way, whether each square starts at this position (it considers first the entire x-axis before moving to the y-axis). Most of the search is in fact spent in searching for the x-coordinates of the squares. The search procedure

```
search {
    forall(p in Positions)
        forall(s in Squares)
            try x[s] = p | x[s] <> p endtry;
    forall(p in Positions)
        forall(s in Squares)
            try y[s] = p | y[s] <> p endtry;
};
```

Name	Duration	Precedences
masonry	7	{}
carpentry	3	{masonry}
roofing	1	{carpentry}
plumbing	8	{masonry}
facade	2	{plumbing, roofing}
windows	1	{roofing}
garden	1	{roofing, plumbing}
ceiling	3	{masonry}
painting	2	{ceiling}
moving	1	{windows, facade, garden, painting}
completion	0	{moving}

Figure 2.7: Activities and Precedences for the House Problem.

captures this strategy. The instruction

```
forall(p in Positions)
    forall(s in Squares)
        try x[s] = p | x[s] <> p endtry;
```

considers all positions along the x-axis. For each such position p, the instruction considers all squares and nondeterministically chooses if the square starts (x[s]=p) or does not start (x[s] <> p) at position p. The instruction

```
forall(p in Positions)
    forall(s in Squares)
        try y[s] = p | y[s] <> p endtry;
```

applies the same principle to the y-axis. This choice strategy, together with the redundant constraints, helps OPL prune the search space substantially. Of course, the choice process considers both the x- and the y-coordinates.

2.3 Scheduling

The other novel aspect of OPL is its support for scheduling applications, which are ubiquitous in industry. The OPL implementation uses specialized algorithms for scheduling applications that can reduce the search space substantially. Scheduling applications in OPL are discussed in Chapter 11; this section merely illustrates some of the concepts on a simple application.

The application is scheduling the activities necessary to build a house. Figure 2.7 depicts the activities, their duration, and the precedence constraints. For instance, facade is an activity of

2. A Short Tour of OPL

duration 2 that can start only when `plumbing` and `roofing` are completed. In addition, each activity costs an amount proportional to its duration. This amount, to be paid at the beginning of the activity, is $1,000 per day. The total budget is of course $29,000. Only $20,000 is available at the beginning of the project; the remaining $9,000 becomes available 15 days thereafter. The goal of the application is of course to minimize the project duration subject to the precedence and budget constraints.

Statement 2.24 depicts an OPL model for this application. The purpose here is not to explain in full detail all concepts in the model, but to give a feeling of the kind of support provided by OPL on scheduling applications. The main concept in scheduling application is the notion of activity, which is the association of two integer variables: a starting date and a duration. The instruction

```
Activity a[t in Tasks](duration[t]);
```

declares an array `a` of activities. The array is indexed by an enumerated set and activity `a[t]` has duration `duration[t]`. The starting dates of these activities are the primary output of the model. The instruction

```
DiscreteResource budget(29000);
```

declares the budget as a discrete resource with capacity 29,000. Activities are able to consume this budget, as becomes clear below. Resources are the second main scheduling concept, and OPL supports a variety of resources, including unary and discrete resources and reservoirs. Instructions of the form

```
a[masonry] precedes a[carpentry];
```

specify the precedence constraints of the application. The constraint specifies that the starting date of activity `carpentry` is greater or equal to the end date of activity `masonry`. The instruction

```
capacityMax(budget,0,15,20000);
```

specifies that the capacity of `budget` is only 20,000 for the first 15 days. The final constraint

```
forall(t in Tasks)
    a[t] consumes(1000*duration[t]) budget;
```

specifies that activity `a[t]` consumes the budget in a quantity proportional to its duration. Note that the objective function is to minimize the end date of the last activity. Here is the optimal solution returned by OPL for Statement 2.24:

```
Optimal Solution with Objective Value:   21
  a[masonry] = [0 -- 7 --> 7]
```

```
enum Tasks
   { masonry, carpentry, plumbing, ceiling, roofing, painting,
     windows,facade,garden,moving};

int duration[Tasks] = [7,3,8,3,1,2,1,2,1,1];

Activity a[t in Tasks](duration[t]);

DiscreteResource budget(29000);

minimize
   a[moving].end
subject to {
   a[masonry] precedes a[carpentry];
   a[masonry] precedes a[plumbing];
   a[masonry] precedes a[ceiling];
   a[carpentry] precedes a[roofing];
   a[ceiling] precedes a[painting];
   a[roofing] precedes a[windows];
   a[roofing] precedes a[facade];
   a[plumbing] precedes a[facade];
   a[roofing] precedes a[garden];
   a[plumbing] precedes a[garden];
   a[windows] precedes a[moving];
   a[facade] precedes a[moving];
   a[garden] precedes a[moving];
   a[painting] precedes a[moving];

   capacityMax(budget,0,15,20000);

   forall(t in Tasks)
      a[t] consumes(1000*duration[t]) budget;
};
```

Statement 2.24: The House Problem (house2.mod).

```
a[carpentry] = [7 -- 3 --> 10]
a[plumbing]  = [7 -- 8 --> 15]
a[ceiling]   = [15 -- 3 --> 18]
a[roofing]   = [10 -- 1 --> 11]
a[painting]  = [18 -- 2 --> 20]
a[windows]   = [11 -- 1 --> 12]
a[facade]    = [15 -- 2 --> 17]
a[garden]    = [15 -- 1 --> 16]
a[moving]    = [20 -- 1 --> 21]
```

It specifies, for instance, that activity `ceiling` starts at time 15, has a duration of 3, and is completed by time 18.

2.4 Notes and References

The Volsay problem is taken from course notes from the Jean Fichefet's 1984 operations research class at the University of Namur. The production and mixing problems are adapted from a similar problem in [23]. The first constraint-programming solution of the magic series was presented in [28]. The first models presented are based on the constraint program presented in [33]. The model with constraint `distribute` is based on [25]. The first constraint program for the stable marriage was presented in [25], the model presented here is significantly simpler. See [12] for a comprehensive discussion of the stable marriage problem. The perfect square problem and its variants have attracted much attention in the constraint-programming community. Colmerauer [4] presents a program to fill a rectangle with squares of different sizes using linear equations, inequalities, and disequations over rational numbers. The sizes of the rectangles are not known and the resulting program is a beautiful constraint program. Aggoun and Beldiceanu [1] present a `CHIP` program that solves the perfect square problem when the sizes of the squares are given. The program uses a specialized cumulative constraint and exploits the link between cumulative constraints and packing in two dimensions. The model described in this chapter was inspired by a program in [30] whose basic idea was suggested by Alain Colmerauer. The house problem is taken from [24].

3 Models

This brief chapter fixes the syntactic conventions used in this book and describes the overall structure of OPL models.

3.1 Syntactic Conventions

The following conventions are used to describe the grammar of OPL: terminal symbols are denoted in typewriter font (e.g., `solve`), $\langle nt \rangle$ denotes a nonterminal symbol nt, [*object*] denotes an optional grammar segment *object*, { *object* } denotes zero, one, or several times the grammar segment *object*, *object*$^+$ denotes an expression *object* { , *object* } while *object** denotes an expression *object* { ; *object* } When a nonterminal symbol, say $\langle n \rangle$, is defined by several rules, say $\langle a \rangle$, $\langle b \rangle$, and $\langle c \rangle$, we use the notation

$$\begin{aligned} \langle n \rangle &\rightarrow \langle a \rangle \\ &\rightarrow \langle b \rangle \\ &\rightarrow \langle c \rangle \end{aligned}$$

or

$$\langle n \rangle \rightarrow \langle a \rangle \mid \langle b \rangle \mid \langle c \rangle$$

depending on convenience in context.

3.2 Terminal Symbols

The basic building blocks of OPL are integers (terminal `Integer`), floating-point numbers (terminal `Float`), identifiers (terminal `Id`), strings (terminal `String`), and the keywords of the language (e.g., `forall`). Identifiers in OPL start with a letter and can contain only letters, digits, and the symbol _. Note that letters in OPL are case-sensitive. Integers are sequences of digits, possibly prefixed by a minus sign. Floats can be described in decimal notation (e.g., `3.4` or `-2.5`) or in scientific notation (e.g., `3.5e-3` or `-3.4e10`). The OPL reserved words are listed in Table 3.1.

3.3 Models

The structure of OPL models is depicted in Figure 3.1. An OPL statement consists of a sequence of declarations, an instruction, a search procedure, a sequence of display instructions, and a sequence of data initializations. The declarations, search procedure, displays, and data initializations are all optional. The following chapters describe each of these components in detail.

Activity	AlternativeResources	DiscreteResource	Reservoir	UnaryResource
all	alldifferent	assert	assignAlternatives	branch
branchLow	branchUp	break	breakOnDuration	breakable
by	capacityMax	capacityMin	circuit	constraint
consumes	data	decreasing	depthfirst	dichotomic
diff	disjunctive	display	distribute	do
edgeFinder	else	endif	endtry	enum
float	forall	generate	generateMax	generateMin
generateSeq	generateSize	generateSym	if	in
increasing	infinity	initialize	int	inter
let	linear	max	maximize	maxint
min	minimize	mod	not	of
onDomain	onRange	onValue	once	ordered
path	periodicBreak	piecewise	precedes	prod
produces	provides	range	rank	rankGlobal
rankLocal	relaxation	requires	scheduleHorizon	scheduleOrigin
search	select	set	setTimes	setting
solve	split	splitLow	splitUp	struct
subject	subset	sum	symdiff	then
to	try	tryRankFirst	tryRankLast	tryall
type	union	using	var	visualize
when	while	with		

Table 3.1: Reserved Words of OPL.

$\langle Model \rangle \rightarrow \underline{\{} \langle Decl \rangle \underline{\}}$
$ \langle Instr \rangle$
$ \underline{[} \langle Search \rangle \underline{]}$
$ \underline{\{} \langle Display \rangle \underline{\}}$
$ \underline{\{} \langle Data \rangle \underline{\}}$

Figure 3.1: The Syntax of Models.

4 Data Modeling

This chapter describes the basic concepts available for modeling data in OPL. Section 4.1 reviews the basic types, i.e., integers, floats, and enumerated values. Section 4.2 describes how these basic types can be combined using arrays, records, and sets to obtain complex data structures. Section 4.3 shows how to declare variables in OPL, Section 4.4 describes specific modeling tools for scheduling applications, Section 4.5 shows how to declare constraints, Section 4.6 discusses assertions to check data consistency, while Section 4.7 describes data initialization. Figure 4.1 describes the high-level syntax of data modeling declarations.

4.1 Basic Data Types

The three basic data types available in OPL are integers, floats, and enumerated values.

4.1.1 Integers

OPL provides the subset of the integers ranging from -2^{31} to 2^{31} as a basic data type. OPL also contains the integer constant `maxint`, which represents the largest positive integer available. A declaration of the form

```
int a = 25;
```

declares an integer `a` whose value is 25. It is also possible to declare natural numbers (i.e., nonnegative integers) using a declaration of the form

```
int+ a = 25;
```

When declaring natural numbers, OPL automatically checks whether the value used in the initialization is nonnegative. An execution error is raised when this is not the case. The initialization of an integer can be specified by an expression. For instance, the declaration

```
int n = 3;
int size = n*n;
```

initializes `size` as the square of `n`. Expressions are covered in detail in Chapter 5. Integers can also be initialized by querying users. A declaration such as

```
int nbQueens << "Number of Queens:";
```

queries users with the message "number of queens:" and expects an integer, which is then used to initialize `nbQueens`.

$\langle Decl \rangle$ → $\langle TypeDecl \rangle$;
→ $\langle DataDecl \rangle$;
→ $\langle VarDecl \rangle$;
→ $\langle ActivityDecl \rangle$;
→ $\langle ResourceDecl \rangle$;
→ $\langle ConstraintDecl \rangle$;
→ $\langle Assert \rangle$;
→ $\langle Initialize \rangle$;
→ $\langle ScheduleInit \rangle$;
→ $\langle SettingDecl \rangle$;

Figure 4.1: The Syntax of Declarations.

4.1.2 Floats

OPL also provides a subset of the real numbers as the basic data type `float`. The implementation of floats in OPL obeys the IEEE 754 standard for floating-point arithmetic and the data type `float` in OPL uses double-precision floating-point numbers. OPL also has a predefined float constant `infinity` to represent the IEEE infinity symbol. Declarations of floats are essentially similar to declarations of integers. The declaration

```
float f = 3.2;
```

declares a float `f` whose value is 3.2. Once again, the value of the float can be specified by an arbitrary expression. Nonnegative floats play a special role in mathematical programming and are supported directly in OPL. For instance, the declaration

```
float+ f = 4.0;
```

declares a nonnegative float `f`. Once again, OPL checks if the initialization actually produces a nonnegative value and raises an error otherwise.

4.1.3 Enumerated Types

Enumerated types are an integral part of many programming languages and help producing more readable programs. Enumerated types in OPL are similar in spirit to those of the programming languages C and PASCAL and are specified by listing their values. For instance, the declaration

```
enum Days {Monday, Tuesday, Wednesday, Thursday, Friday, Saturday, Sunday};
```

4. Data Modeling

declares the enumerated type `Days` that consists of the values `Monday, Tuesday, Wednesday, Thursday, Friday, Saturday,` and `Sunday.` The names in an enumerated type have a global scope and cannot be used in any other enumerated type. Once an enumerated type `T` is declared, it is possible to declare data of type `T`. For instance, the declaration

```
Days myFavoriteDay = Sunday;
```

declares a data `myFavoriteDay` of type `Days` and initializes it to `Sunday`. Of course, it is possible to define arrays and sets of type `Days`, as shown later in this chapter.

4.2 Data Structures

More complex data types can be built from the basic data types. This section gives an overview of the basic data structures available in OPL, i.e., ranges, arrays, records, and sets.

4.2.1 Ranges

Integer ranges are fundamental in OPL, since they are often used in arrays and variable declarations, as well as in aggregate operators, queries, and quantifiers. An integer range is specified by giving its lower and upper bounds, as in

```
range Rows 1..10;
```

which declares a range `1..10`. The lower and upper bounds can also be given by expressions, as in

```
int n = 8;
range Rows [n+1..2*n+1];
```

Note that ranges are normally given in between brackets. If the lower bound is a number (i.e., an integer or a float), then the brackets can be omitted. Once a range has been defined, it is possible to declare data taking their values in this range. For intance, the declarations

```
range R 1..10;
R r = 8;
R v[1..3] = [1, 2, 3];
```

defines an integer `r` ranging in `R` and an array whose elements range in `R`. Once again, OPL checks whether the initialization produces a value inside the range and raises an error otherwise. OPL also supports the declaration of float ranges that specify a subset of the reals. For intance,

```
range float fr 1.2..2.2;
```

specifies the subset of the reals in the interval `[1.2,2.2]`.

4.2.2 Arrays

Arrays are, of course, fundamental in many applications; this section reviews the support for arrays available in OPL.

4.2.2.1 One-Dimensional Arrays

One-dimensional arrays are the simplest arrays in OPL and vary according to the type of their elements and index sets. A declaration of the form

```
int a[1..4] = [10, 20, 30, 40];
```

declares an array of four integers `a[1],...,a[4]` whose values are 10, 20, 30, and 40. It is of course possible to define arrays of other basic types. For instance, the instructions

```
int+ a[1..4] = [10, 20, 30, 40];
float f[1..4] = [1.2, 2.3, 3.4, 4.5];
Days d[1..2] = [Monday, Wednesday];
```

declare arrays of natural numbers, floats, and enumerated values, respectively.

The index sets of arrays in OPL are very general and can be integer ranges, enumerated types, and arbitrary finite sets. In the examples so far, index sets were given explicitly, but it is possible to use a previously defined range, as in

```
range R 1..4;
int a[R] = [10, 20, 30, 40];
```

The declaration

```
int a[Days] = [10, 20, 30, 40, 50, 60, 70];
```

describes an array indexed by an enumerated type; its elements are `a[Monday],...,a[Sunday]`.

Arrays can also be indexed by finite sets of arbitrary types. This feature is fundamental in OPL to exploit sparsity in large linear programming applications, as discussed in detail in Chapter 9. For instance, the declaration

```
struct Edge {
    int orig;
    int dest;
};
{Edge} Edges = {<1,2>, <1,4>, <1,5>};
int a[Edges] = [10, 20, 30];
```

defines an integer array `a` indexed by a finite set of records; its elements are `a[<1,2>]`, `a[<1,4>]`, and `a[<1,5>]`. Records are described in detail in Section 4.2.3.

4.2.2.2 Multidimensional Arrays

OPL supports the declaration of multidimensional arrays. For instance, the declaration

```
int a[1..2,1..3] = ...;
```

declares a two-dimensional array indexed by two integer ranges. Indexed sets of different types can of course be combined, as in

```
int a[Days,1..3] = ...;
```

which is a two-dimensional array whose elements are of the form `a[Monday,1]`.

It is interesting to contrast multidimensional and one-dimensional arrays of records. Consider the declaration

```
enum Warehouses ...;
enum Customers ...;
int transp[Warehouses,Customers] = ...;
```

which declares a two-dimensional array **transp**. This array may represent the units shipped from a warehouse to a customer. In large-scale applications, it is likely that a given warehouse delivers only to a subset of the customers. The array **transp** is thus likely to be sparse, i.e., it will contain many zero values. The sparsity can be exploited by declarations of the form

```
enum Warehouses ...;
enum Customers ...;
struct Route {
    Warehouses w;
    Customers c;
};
{Route} routes = ...;
int transp[routes] = ...;
```

This declaration specifies a set **routes** that contains only the relevant pairs (warehouse,customer). The array **transp** can then be indexed by this set, exploiting the sparsity present in the application. It should be clear that, for large-scale applications, this approach leads to substantial reductions in memory consumption.

4.2.2.3 Initializing Arrays

There are various ways of initializing arrays, and data in general, in OPL. This section reviews the basic principles for arrays. Initializations are covered in more detail in Section 4.7. An array can be initialized by listing its values, as in most of the examples presented so far. Multidimensional arrays in OPL are, in fact, arrays of arrays and must be initialized accordingly. For instance, the declaration

```
int a[1..2,1..3] = [
    [10, 20, 30],
    [40, 50, 60]
];
```

initializes a two-dimensional array by giving initializations for the one-dimensional arrays of its first dimension. It is easy to see how to generalize this initialization to any number of dimensions. An array can also be initialized by specifying pairs (index,value), as in the declaration

```
int a[Days] = #[
    Monday:    1,
    Tuesday:   2,
    Wednesday: 3,
    Thursday:  4,
    Friday:    5,
    Saturday:  6,
    Sunday:    7
]#;
```

There are a couple of remarks to be made here. First, when the initialization is specified by (index,value) pairs, the delimiters #[and]# must be used instead of [and]. Second, the ordering of the pairs can be arbitrary. Of course, OPL verifies that each entry in the array is initialized exactly once. These two forms of initializations can be combined arbitrarily, as in

```
int a[1..2,1..3] = #[
    2: [40, 50, 60],
    1: [10, 20, 30]
]#;
```

The last three forms of initialization are discussed at length in Section 4.7 and are presented here merely for completeness. Arrays can also be initialized in a **data** instruction, in which case the declaration has the form

```
int a[1..2,1..3] = ...;
```

and the actual initialization is given in a **data** instruction or in a separate file in OPL STUDIO. Arrays can also be initialized by an **initialize** instruction, as in

4. Data Modeling

```
int a[1..8];
initialize
    forall(i in 1..8)
        a[i] = i + 1;
```

which initializes the array in such a way that `a[1]` = 2, `a[2]` = 3, and so on. Finally, arrays can be initialized by specifying a data file containing the values of the array, as in

```
int a[1..80000] < "a.dat";
```

This last form of initialization is probably most appropriate for large data files, since it is more memory-efficient. Its format is described later in this chapter.

4.2.2.4 Generic Arrays

OPL also supports generic arrays, i.e, arrays whose elements are initialized by a generic expression. These generic arrays may significantly simplify the modeling of an application. The declaration

```
int a[i in 1..10] = i+1;
```

declares an array of 10 elements such that the value of `a[i]` is `i+1`. Generic arrays can of course be multidimensional, as in

```
int m[i in 0..10, j in 0..10] = 10*i + j;
```

which initializes element `m[i,j]` to `10*i + j`. Generic arrays are useful in performing some simple transformations. For instance, generic arrays can be used to transpose matrices in a simple way, as in

```
int m[Dim1,Dim2] = ...;
int t[j in Dim2,i in Dim1] = m[i,j];
```

More generally speaking, generic arrays can be used to permute the indices of arrays in simple ways. Note that generic arrays are somewhat redundant with **initialize** instructions. However, it is recommended to use generic arrays when possible, since they make the model more explicit and readable.

4.2.3 Records

As shown previously, data structures in OPL can also be constructed using records that cluster together closely related data. The definition of records in OPL closely follows the syntax of records in C. For instance, the declaration

```
struct Point {
    int x;
    int y;
};
```

declares a record `Point` consisting of two fields `x` and `y` of type integer. Once a record type `T` has been declared, records, arrays of records, set of records of type `T`, records of records can be declared, as in

```
Point p = <2,3>;
Point point[1..3] = [<1,2>, <2,3>, <3,4>];
{Point} points = {<1,2>, <2,3>};
struct Rectangle {
    Point ll;
    Point ur;
};
```

These declarations respectively declare a point, an array of three points, a set of two points, and a record type where the fields are points. The various fields of a record can be accessed in the traditional way by suffixing the record name with a dot and the field name, as in

```
Point p = <2,3>;
int x = p.x;
```

which initializes `x` to the field `x` of record `p`. Note that the field names are local to the scope of the records. Records are initialized by giving the list of the values of the various fields, as in

```
Point p = <2,3>;
```

which initializes `p.x` to 2 and `p.y` to 3. They can also be initialized by listing <field name, value> pairs, as in

```
Point p = #<y:3, x:2>#
```

As with arrays, the delimiters < and > are replaced by #< and ># and the ordering of the pairs is not important. Once again, OPL checks if all fields are initialized exactly once. The type of the fields can be arbitrary and they can contain arrays and sets. For instance, the declaration

```
struct Rectangle {
    int id;
    Point p[2];
};
```

declares a record with two fields, the first an integer and the second an array of two points. A specific "rectangle" can be declared as

```
Rectangle r = <1, [<0,0>, <10,10>]>;
```

The declaration

```
enum Task ...;
struct Precedence {
    Task name;
    {Task} after;
};
```

defines a record whose first field is an enumerated value and whose second field is a set of enumerated values. A possible "precedence" can be declared as follows:

```
Precedence p = <a1, {a2, a3, a4, a5}>;
```

assuming that a1,..,a5 are elements of Task.

4.2.4 Sets

OPL also supports sets of arbitrary types to model data in applications. If T is a type, then {T} denotes the type "set of T". For instance, the declaration

```
{int} setInt = ...;
{Precedence} precedences = ...;
```

declares a set of integers and a set of precedences. Sets can be initialized in various ways. The simplest way to initialize a set is by listing its values explicitly. For instance, the declaration

```
struct Precedence {
    int before;
    int after;
};
{Precedence} precedences = {<1,2>, <1,3>, <3,4>};
```

initializes a set of records explicitly. The instruction

```
struct Precedence {
    int before;
    int after;
};
```

```
{Precedence} precedences = {
    #<before :  1, after :   2>#;
    #<before :  2, after :   3>#;
    #<before :  3, after :   4>#
};
```

has of course a similar effect. Sets can also be initialized by set expressions using previously defined sets and operations such as union, intersection, difference, and symmetric difference.[1], which are described in Chapter 5. For instance, the declarations

```
{int} s1 = {1,2,3};
{int} s2 = {1,4,5};
{int} i = s1 inter s2;
{int} u = s1 union s2;
{int} d = s1 diff s2;
{int} sd = s1 symdiff s2;
```

initializes i to {1}, u to {1,2,3,4,5}, d to {2,3}, and sd to {2,3,4,5}. Note that explicitly defined sets (e.g., {1,2,3}) cannot appear in these expressions.

It is also possible to initialize a set from a range expression. For instance, the declaration

```
{int} s = 1..10
```

initializes s to {1,2,..,10}, while the expression

```
{int} s = 1..10 by 3;
```

initializes s to {1,4,7,10}. It is important to point out at this point that sets initialized by ranges are represented explictly (unlike ranges). As a consequence, a declaration of the form

```
{int} = s = 1..100000;
```

creates a set where all the values 1, 2, ..., 100000 are explicitly represented, while the range

```
range s 1..100000;
```

represents only the bounds explicitly.

Sets can also be specified by queries, which resemble typical relational database queries. For instance, the declaration

```
{int} s = {i | i in 1..10 :  i mod 3 = 1};
```

---

[1] The symmetric difference of two sets A and B is $(A \cup B) \setminus (A \cap B)$.

4. Data Modeling

initializes s with the set $\{1,4,7,10\}$. A query is a conjunction of expressions of the form

```
p in S : condition
```

where p is a parameter (or a tuple of parameters), S is a range, an enumerated type, or a finite set, and condition is an expression. These expressions are also used in forall statements and aggregate operators and are discussed in detail in Chapter 6. The declaration

```
enum Resources ...;
enum Tasks ...;
Tasks res[Resources] = ...;
struct Disjunction {
    Task first;
    Task second;
};
Disjunction disj = {#<first:  i, second:  j># |
    r in Resources & ordered i,j in res[r]
};
```

is a more interesting example, showing a conjunction of expressions, and is explained in detail in Chapter 6. Queries are often useful in transforming a data structure (e.g., the data stored in a file) into a data structure more appropriate for stating the model effectively. Consider, for instance, the declarations

```
enum Nodes ...;
range Bool 0..1;
Bool edges[Nodes,Nodes] = ...;
```

which describe the edges of a graph in terms of a Boolean adjacency matrix. It may be important for the model to use a sparse representation of the edges (because, for instance, edges are used to index an array). The declaration

```
struct Edge {
    Nodes o;
    Nodes d;
};
{Edge} setEdges = {<o,d> | o,d in Nodes :  edges[o,d]};
```

computes this sparse representation using a simple query. It is of course possible to define generic arrays of sets. For instance, the declaration

```
{int} a[i in 3..4] = {e | e in 1..10 :  e mod i = 0};
```

initializes a[3] to $\{3,6,9\}$ and a[4] to $\{4,8\}$.

⟨*TypeDecl*⟩	→	`enum` ⟨*Id*⟩ ⟨*EnumDef*⟩⁺
	→	`range` ⟨*RangeDecl*⟩⁺
	→	`set` ⟨*SetDecl*⟩⁺
	→	`struct` ⟨*StructDecl*⟩⁺
⟨*EnumDecl*⟩	→	`{` ⟨*Id*⟩⁺ `}`
	→	...
	→	`< String`
⟨*RangeDecl*⟩	→	⟨*Id*⟩ ⟨*Range*⟩
	→	`float` ⟨*Id*⟩ ⟨*FloatRange*⟩
⟨*Range*⟩	→	`[` ⟨*Expr*⟩ `..` ⟨*Expr*⟩ `]`
	→	`Integer ..` ⟨*Expr*⟩
⟨*FloatRange*⟩	→	`[` ⟨*Expr*⟩ `..` ⟨*Expr*⟩ `]`
	→	`Integer ..` ⟨*Expr*⟩
	→	`Float ..` ⟨*Expr*⟩
⟨*SetDecl*⟩	→	⟨*Id*⟩ ⟨*Id*⟩
⟨*StructDecl*⟩	→	⟨*Id*⟩ `{` ⟨*Field*⟩* `}`
⟨*Field*⟩	→	⟨*Type*⟩ ⟨*Id*⟩ ⟨*Subscripts*⟩

Figure 4.2: The Syntax of Type Declarations.

4.2.5 Summary of the Syntax

Figure 4.2 summarizes the syntax of type declarations, including the declaration of enumerated sets, ranges, and records. Figure 4.3 summarizes the syntax of data declarations.

4.3 Variables

The purpose of an OPL model is to find values for the variables such that all constraints are satisfied or, in optimization problems, to find values for the variables that satisfy all constraints and optimize a specific objective function. Variables in OPL are thus essentially mathematical variables and differ fundamentally from variables in programming languages such as C and PASCAL. A variable declaration in OPL specifies the type and set of possible values for the variable. Figure 4.4 summarizes the syntax of variable declarations.

Once again, variables can be of different types (integer, float, enumerated types) and it is possible to define multidimentional arrays of variables. The declaration

4. Data Modeling

⟨*DataDecl*⟩ → ⟨*Type*⟩ ⟨*Id*⟩ ⟨*Subscripts*⟩ ⟨*DataInit*⟩
⟨*Type*⟩ → float | float+ | int | int+ | ⟨*Id*⟩ | { ⟨*Type*⟩ }
⟨*Subscripts*⟩ →
 → [*Subscript*⁺]
⟨*Subscript*⟩ → ⟨*Range*⟩
 → ⟨*Id*⟩
 → ⟨*Id*⟩ in ⟨*Id*⟩
 → ⟨*Id*⟩ in ⟨*Range*⟩

Figure 4.3: The Syntax of Data Declarations.

```
var int transp[Orig,Dest] in 0..100;
```

declares a two-dimensional array of integer variables. The variables are constrained to take their values in the range 0..100; i.e., any solution to the model containing this declaration must assign values between 0 and 100 to these variables. Arrays of variables can be constructed using the same index sets as arrays of data. In particular, it is also possible, and desirable for larger problems, to index arrays of variables by finite sets. For instance, the excerpt

```
struct Route {
    City orig;
    City dest;
};
{Route} routes = ...;
var int transp[routes] in 0..100;
```

declares an array of variables **transp** that is indexed by the finite set of records **routes**. It is also possible to use genericity for initializing the domain of the variables. For instance, the excerpt

```
struct Route {
    City orig;
    City dest;
};
{Route} routes = ...;
int capacity[routes] = ...;
var int transp[r in routes] in 0..capacity[r];
```

declares an array of variables indexed by the finite set `routes` such that variable `transp[r]` ranges over `0..capacity[r]`. The array `capacity` is also indexed by the finite set `routes`. Note also that variables can be declared to range over a user-defined range. For instance, the excerpt

```
range Capacity 0..limitCapacity;
var Capacity transp[Orig,Dest];
```

declares an array of integer variables ranging over `Capacity`. Variables can of course be declared individually, as in

```
var int averageDelay in 0..maxDelay;
```

As mentioned previously, variables can be of different types. The excerpt

```
var float+ transp[orig, dest];
```

declares a two-dimensional array of float variables that are constrained to be non-negative (i.e., their ranges is $0..\infty$). Float variables can be assigned more specific ranges, as in

```
var float+ transp[o in Orig, d in Dest] in 0..cap[o,d];
```

which declares a two-dimensional array of float variables, where variable `transp[o,d]` ranges over the set `0..cap[o,d]`. This last declaration is of course equivalent to the declaration

```
var float transp[o in Orig, d in Dest] in 0..cap[o,d];
```

Variables over enumerated types are also useful for a variety of applications. The excerpt

```
enum Men ...;
enum Women ...:
var Women wife[Men];
```

declares an array `wife` of variables, indexed by the enumerated type `Men` and of type `Women`. Every solution to the model containing this excerpt assigns an element of `Women` to each variable in the array `wife`.

4.4 Data Types for Scheduling Applications

OPL also provides a number of data types tailored for scheduling applications. These data types represent important abstractions in modeling scheduling problems and OPL exploits them in special-purpose constraint-solving algorithms. The syntax of scheduling declarations is given in Figure 4.5.

4. Data Modeling

$\langle VarDecl \rangle \rightarrow$ `var` $\langle VarDeclSfx \rangle^+$
$\langle VarDeclSfx \rangle \rightarrow \langle Id \rangle \langle Id \rangle \langle Subscripts \rangle$
$ \rightarrow$ `int` $\langle Id \rangle \langle Subscripts \rangle$
$ \rightarrow$ `int` $\langle Id \rangle \langle Subscripts \rangle$ `in` $\langle Expr \rangle$ `..` $\langle Expr \rangle$
$ \rightarrow$ `int+` $\langle Id \rangle \langle Subscripts \rangle$
$ \rightarrow$ `int+` $\langle Id \rangle \langle Subscripts \rangle$ `in` $\langle Expr \rangle$ `..` $\langle Expr \rangle$
$ \rightarrow$ `float` $\langle Id \rangle \langle Subscripts \rangle$
$ \rightarrow$ `float` $\langle Id \rangle \langle Subscripts \rangle$ `in` $\langle Expr \rangle$ `..` $\langle Expr \rangle$
$ \rightarrow$ `float+` $\langle Id \rangle \langle Subscripts \rangle$
$ \rightarrow$ `float+` $\langle Id \rangle \langle Subscripts \rangle$ `in` $\langle Expr \rangle$ `..` $\langle Expr \rangle$

Figure 4.4: The Syntax of Variable Declarations.

$\langle ScheduleInit \rangle \rightarrow$ `scheduleHorizon =` $\langle Expr \rangle$
$ \rightarrow$ `scheduleOrigin =` $\langle Expr \rangle$
$\langle ActivityDecl \rangle \rightarrow$ `Activity` $\langle Id \rangle \langle Subscripts \rangle$ [$\langle Breakable \rangle$]
$ \rightarrow$ `Activity` $\langle Id \rangle \langle Subscripts \rangle$ ($\langle Expr \rangle$) [$\langle Breakable \rangle$]
$\langle Breakable \rangle \rightarrow$ `breakable` [`if` $\langle Relation \rangle$]

$\langle ResourceDecl \rangle \rightarrow$ `UnaryResource` $\langle Id \rangle \langle Subscripts \rangle$
$ \rightarrow$ `DiscreteResource` $\langle Id \rangle \langle Subscripts \rangle$ ($\langle Expr \rangle$)
$ \rightarrow$ `DiscreteResource` $\langle Id \rangle \langle Subscripts \rangle$ ($\langle Expr \rangle$) `using disjunctive`
$ \rightarrow$ `DiscreteResource` $\langle Id \rangle \langle Subscripts \rangle$ ($\langle Expr \rangle$) `using edgeFinder`
$ \rightarrow$ `Reservoir` $\langle Id \rangle \langle Subscripts \rangle$ ($\langle Expr \rangle$)
$ \rightarrow$ `Reservoir` $\langle Id \rangle \langle Subscripts \rangle$ ($\langle Expr \rangle$, $\langle Expr \rangle$)
$ \rightarrow$ `AlternativeResources` $\langle Id \rangle \langle Subscripts \rangle$ ($\langle Composite \rangle$)

Figure 4.5: The Syntax of Scheduling Declarations.

4.4.1 Origin and Horizon

All scheduling concepts used in OPL are defined over a global time interval

[scheduleOrigin, scheduleHorizon)

closed on the left and open on the right, like all time intervals in OPL. OPL has a default value for both the origin (0) and the horizon (a large number). However, it is recommended that they be specified for particular applications, since smaller time intervals make OPL more space- and time-efficient. The global origin and horizon can be specified by using instructions of the form

```
scheduleOrigin = 0;
scheduleHorizon = 365;
```

It is of course possible to use expressions to initialize these values, as in

```
scheduleHorizon = sum(t in Tasks) duration[t];
```

Note that the declarations of `scheduleOrigin` and `scheduleHorizon`, if present, must come before the declarations of any other scheduling objects in the model. This is consistent with our convention of requiring that each object be used only after it is defined.

4.4.2 Activities

Probably the most fundamental concept in OPL for scheduling applications is the *activity*. An activity can be thought of as an object containing three data items, a starting date, a duration, and an ending date, together with the *duration constraint* that the ending date is the starting date plus the duration. In many applications, the duration of an activity is known and the activity is declared together with its duration, as in

```
Activity carpentry(10);
```

which declares an activity `carpentry` whose duration is 10. The starting and ending dates of `carpentry` are integer variables taking their values in the global time interval and consistent with the duration constraint. Activities can also be given a variable duration, in which case one commonly declares the task with an integer variable representing the duration, as in

```
var int durationCarpentry in 8..10;
Activity carpentry(durationCarpentry);
```

which declares an activity `carpentry` whose duration is between 8 and 10. The duration variable `durationCarpentry` can appear in the problem constraints and the possible values for duration may thus be further constrained. It is also possible to declare an activity without specifying its duration, as in

```
Activity carpentry;
```
in which case the duration is an integer variable ranging over the interval

$[0, \mathtt{scheduleHorizon} - \mathtt{scheduleOrigin}]$.

Arrays of activities can be declared in the usual way and the durations may be specified as described previously. For instance, it is usual to declare an array of activities as follows:

```
Activity tasks[t in 1..10](duration[t]);
```

The statement declares an array of 10 activities whose durations are `duration[1],...,duration[10]` respectively. The starting date, the duration, and the ending date of an activity are accessed in the way as the fields of a structure. For instance, `carpentry.start`, `carpentry.end`, and `carpentry.duration` represent the starting date, the ending date, and the duration of activity `carpentry`.

When modeling real applications, it may be important to recognize periods, such as weekends, when no activity can be scheduled. In OPL, these periods are called *breaks*. OPL provides several tools to specify breaks; these are discussed in Chapter 5. Some activities can be interrupted by breaks, while others cannot. In OPL, activities that can be interrupted by breaks are called *breakable activities*: they can start before a break and be resumed after a break. Activities in OPL are assumed to be unbreakable unless specified otherwise. The declaration

```
Activity a breakable;
```

specifies that `a` is a breakable activity. Of course, breakable activities behave very much like other activities: they have a starting date, an ending date, and a duration, and they may require resources like activities. Their only added functionality is the ability to be interrupted by breaks. It is possible to declare arrays of breakable activities. For instance, the declaration

```
Activity tasks[t in 1..10](duration[t]) breakable;
```

defines an array of ten breakable activities. Finally, it may be useful to declare an array that has both standard and breakable activities. This makes it easy to produce generic models that are parametrized by the status of each activities. The declaration

```
Activity tasks[t in 1..10](duration[t]) breakable if t in breakableSet;
```

defines an array of 10 activities, some of which can be breakable.

4.4.3 Resources

Resources are a fundamental concept in scheduling applications and it is thus not surprising to find them as basic data types in OPL. OPL in fact supports a variety of resources, including unary and discrete resources, as well as reservoirs.

Unary Resources A unary resource is a resource that cannot be shared by two activities: i.e., as soon as an activity requires the resource, no other activities can make use of that resource. Unary resources are often used to model machines in job-shop scheduling or individual resources in resource-allocation problems. Activities generally require one or several unary resources during their execution. For instance, a job may require both a machine and an operator, both of whom can be modeled by unary resources. Unary resources are declared in OPL as

```
UnaryResource crane;
```

As usual, it is possible to declare arrays of unary resources, as in

```
UnaryResource machines[1..10];
```

which declares an array of 10 unary resources.

Discrete Resources Discrete resources are another fundamental concept offered by OPL for scheduling applications. They are used to model resources that are available in multiple units, all units being considered equivalent and interchangeable as far as the application is concerned. For instance, a discrete resource may be used to model a budget or a pool of identical tools. Activities can then consume the discrete resource (e.g., spend part of the budget) or can require the resource during their execution (e.g., an activity can require a hammer) and return it to the pool upon completion. Discrete resources are declared in OPL by specifying their capacity, as in

```
DiscreteResource budget(30000);
```

which specifies that **budget** is a resource of capacity 30,000, i.e., there is $30,000 is available. Of course, it is possible in OPL to declare arrays of discrete resources, as in

```
DiscreteResource res[t in 1..10](cap[t]);
```

which declares an array of 10 resources, the capacity of resource **t** being **cap[t]**. Note the generic way of specifying the capacity of the discrete resources. If all the discrete resources have the same capacity, say 3, then the declaration

```
DiscreteResource res[1..10](3);
```

is perfectly appropriate. A unary resource is a discrete resource of capacity 1 but the algorithms for unary resources are optimized to exploit all properties of this special case.

Reservoirs Unary and discrete resources are appropriate modeling tools when activities only require or consume resources. Some applications, however, may have a combination of activities, some of which require resources and others of which provide resources. For instance, an activity may require 10 gallons of oil, while another activity may supply oil at some point in time. Resources that can be required (consumed) and provided (produced) are called *reservoirs* in OPL. Reservoirs are declared, as in

```
Reservoir tank1(1000);
Reservoir tank2(1000,100);
```

which declares a reservoir `tank1` whose maximum capacity is 1,000 and initial capacity is 0 and a reservoir `tank2` with the same maximum capacity but with an initial capacity of 100. As usual, it is possible in OPL to declare arrays of reservoirs.

Alternative Resources In many applications, an activity may require one of several resources, that are equivalent from the activity standpoint. However, the resources can sometimes not be equivalent from an application standpoint (e.g., some activities can only be performed by some of these resources). OPL supports these applications by providing the concept of *alternative resources*. Alternative resources are in fact sets of unary resources that can be declared, as in

```
UnaryResources oven[1..10];
AlternativeResources s(oven);
```

This declaration simply specifies that `s` is the set of unary resources `oven[1]`, ..., `oven[10]`. There is a fundamental difference between the above alternative resources and a discrete resource such as

```
DiscreteResource oven(10);
```

If the application is modeled with a discrete oven resource, a task may change ovens during its execution. The only requirement for discrete resources is that, at any time, there are enough ovens for all tasks requiring one. To the contrary, with alternative resources, the task is guaranteed to stay in the same oven during its execution. When permitted by the semantics of the application, it is preferable to use discrete resources rather than alternative resources, since the constraint-solving algorithms are more effective for discrete resources. Sections 11.4.4 and 11.6 discuss on this topic further.

4.5 Constraint Declarations

OPL also allows constraint declarations. This feature is useful in many problems for displaying constraints and analyzing the results of OPL, as well as for removing and restoring constraints in

⟨*ConstraintDecl*⟩ → constraint ⟨*Id*⟩ ⟨*Subscripts*⟩

Figure 4.6: The Syntax of Constraint Declarations.

⟨*Assert*⟩ → assert ⟨*Cstr*⟩

Figure 4.7: The Syntax of Assertions.

OPL STUDIO. The syntax of constraint declarations is shown in Figure 4.6. For instance, a declaration

```
constraint capCstr[Machines];
```

declares an array `capCstr` of constraints. Chapter 5 describes how to initialize elements of this array.

4.6 Data Consistency

OPL also provides assertions to verify the consistency of the model data. This functionality is often useful to avoid tedious model debugging or wrong results due to incorrect input data. These assertions are similar to the `assert` instruction in C. In their simplest form, assertions are simply Boolean expressions that must be true; they raise an execution error otherwise. For instance, it is common in some transportation problems to require that demand matches the supply. The declaration

```
{int} demand[Customers] = ...;
{int} supply[Suppliers] = ...;
```

```
assert sum(s in Suppliers) supply[s] = sum(c in Customers) demand[c];
```

makes sure that the total supply of the suppliers meets the total demand of the customers. This assertion can be generalized to the case of multiple products, as in

```
{int} demand[Customers,Products] = ...;
{int} supply[Suppliers,Products] = ...;
```

```
assert
   forall(p in Products)
      sum(s in Suppliers) supply[s,p] = sum(c in Customers) demand[c,p];
```

which verifies that the total supply meets the total demand for each product. The use of assertions is highly recommended, since they make it possible to detect errors in the data input early, avoiding tedious inspection of the model data and results. Figure 4.7 summarizes the syntax of assertions.

4.7 Initialization

As mentioned previously, there are various ways to initialize data in OPL:

- *inline initialization*: the initialization is given together along the declaration;
- *offline initialization*: the initialization is given subsequently in a `data` instruction or in a separate OPL STUDIO file;
- *computed initialization*: the initialization is given by an `initialize` instruction;
- *file initialization*: the initialization is given in a file.

4.7.1 Inline Initialization

Inline initializations specify the initialization at the same time as the declaration. Inline initializations can contain expressions to initialize data items, such as

```
int a[1..5] = [b+1,b+2,b+3,b+4,b+5];
```

Figure 4.8 summarizes the syntax of inline initializations.

4.7.2 Offline Initialization

Offline initializations specify the initialization of a data item in a `data` instruction. The fact that the data is initialized in a `data` instruction is specified using = ... in the declaration. This functionality makes it possible to have model and data files in OPL STUDIO, each data file corresponding in fact to a `data` instruction. For instance, the excerpt

```
int a[1..2] = ...;
int b[1..3] = ...;
...
data {
    b = [1,2,3];
    a = [3,4];
};
```

⟨*DataInit*⟩	→	
	→	= ...
	→	= ⟨*EltInit*⟩
	→	<< String
	→	< String
⟨*EltInit*⟩	→	< ⟨*EltInit*⟩⁺ >
	→	#< ⟨*NtupleInit*⟩⁺ >#
	→	[⟨*EltInit*⟩⁺]
	→	#[⟨*NEltInit*⟩⁺]#
	→	#{ ⟨*ListSeq*⟩ }#
	→	{}
	→	{ ⟨*EltInit*⟩ | ⟨*ListParameter*⟩ }
	→	{ ⟨*EltInit*⟩⁺ }
	→	⟨*Expr*⟩
	→	⟨*Expr*⟩ .. ⟨*Expr*⟩
	→	⟨*Expr*⟩ .. ⟨*Expr*⟩ by ⟨*Expr*⟩
⟨*NEltInit*⟩	→	⟨*EltInit*⟩ : ⟨*EltInit*⟩
⟨*NtupleInit*⟩	→	⟨*Id*⟩ : ⟨*EltInit*⟩

Figure 4.8: The Syntax of Inline Initializations.

declares two arrays whose initializations are given in the **data** instruction. Note that the order of the initializations in the **data** instruction is not significant and that commas can be omitted as separators in arrays, sets, and records. However, we stress that **data** instructions cannot contain expressions, since they are intended to specify data. The **data** instruction also makes it possible to specify sets of records in very compact ways. Consider the types

```
enum Product {flour, wheat, sugar};
enum City {Providence, Boston, Mansfield};
struct Ship {
    City orig:
    City dest;
    Product p:
};
```

and assume that a set of shipments must be initialized with the values

```
<Providence, Boston, wheat>;
<Providence, Boston, flour>;
<Providence, Boston, sugar>;
<Providence, Mansfield, wheat>;
<Providence, Mansfield, flour>;
<Boston, Providence, sugar>;
<Boston, Providence, flour>;
```

The OPL declaration

```
{Ship} shipment = #{
    <Providence> ::  {<Boston>    ::  {<wheat> <flour> <sugar>}
                     <Mansfield>  ::  {<wheat> <flour>}}
    <Boston>     ::  {<Providence> ::  {<sugar> <flour>}}
}#;
```

factors redundancies across the tuples. The compact form is enclosed in #{ and }# instead of {and }. In addition, it uses a concatenation operator $r :: S$ that, given a tuple r and a set of tuples $S = \{r_1, \ldots, r_n\}$, returns the set of records $r'_1, \ldots, r'_n$, where r'_i is the concatenation of the tuple r and r_i. Figure 4.8 summarizes the syntax of offline initializations.

4.7.3 Computed Initializations

Computed initializations are often useful to transform data stored in some format into a form more appropriate to state the model effectively. Consider a transportation problem in which the input

⟨*Data*⟩	→	`data {` { ⟨*ListInit*⟩ } `} ;`
⟨*InitStatement*⟩	→	⟨*Id*⟩ `=` ⟨*Define*⟩ `;`
⟨*Define*⟩	→	`Integer`
	→	`Float`
	→	`Id`
	→	`<` ⟨*ListDefine*⟩ `>`
	→	`#<` ⟨*ListNTDefine*⟩ `>#`
	→	`{` ⟨*ListDefine*⟩ `}`
	→	`[` ⟨*ListDefine*⟩ `]`
	→	`#[` ⟨*ListNDefine*⟩ `]#`
	→	`#{` ⟨*ListSeq*⟩ `}#`
⟨*ListDefine*⟩	→	
	→	⟨*Define*⟩ ⟨*ListDefine*⟩
	→	⟨*Define*⟩ `,` ⟨*ListDefine*⟩
⟨*NDefine*⟩	→	⟨*Define*⟩ `:` ⟨*Define*⟩
⟨*ListNDefine*⟩	→	
	→	⟨*NDefine*⟩ ⟨*ListNDefine*⟩
	→	⟨*NDefine*⟩ `,` ⟨*ListNDefine*⟩
⟨*NTDefine*⟩	→	`Id :` ⟨*Define*⟩
⟨*ListNTDefine*⟩	→	
	→	⟨*NTDefine*⟩ ⟨*ListNTDefine*⟩
	→	⟨*NTDefine*⟩ `,` ⟨*ListNTDefine*⟩
⟨*ListSeq*⟩	→	⟨*ListSeqo*⟩ `|` ⟨*ListSeqn*⟩
⟨*ListSeqo*⟩	→	⟨*Seqo*⟩
	→	⟨*Seqo*⟩ ⟨*ListSeqo*⟩
	→	⟨*Seqo*⟩ `,` ⟨*ListSeqo*⟩
⟨*Seqo*⟩	→	`<` ⟨*ListDefine*⟩ `>`
	→	`<` ⟨*ListDefine*⟩ `> :: {` ⟨*ListSeqo*⟩ `}`
⟨*ListSeqn*⟩	→	⟨*Seqn*⟩
⟨*ListSeqn*⟩	→	⟨*Seqn*⟩ ⟨*ListSeqn*⟩
	→	⟨*Seqn*⟩ `,` ⟨*ListSeqn*⟩
⟨*Seqn*⟩	→	`#<` ⟨*ListNTDefine*⟩ `>#`
	→	`#<` ⟨*ListNTDefine*⟩ `># :: {` ⟨*ListSeqn*⟩ `}`

Figure 4.9: The Syntax of Offline Initializations.

4. Data Modeling

data is given by a list of tuples of the form

`<orig,dest,c>`

This indicates that there is a route between cities `orig` and `dest` with a shipment cost `c`. For instance, the input data can be described by the instructions

```
struct InputData {
    City orig;
    City dest;
    float c;
};
{InputData} inData = ...:
```

It may be useful in the model to have an array `cost` declared as follows:

```
struct Route {
    City orig;
    City dest;
};
{Route} routes = {<orig,dest> | <orig,dest,c> in inData};
float+ cost[routes];
```

This array can be initialized by the `initialize` instruction

```
forall(d in inData)
    cost[<d.orig,d.dest>] = d.cost;
```

or, equivalently, using tuples as

```
initialize
    forall(<o,d,c> in inData)
        cost[<o,d>] = c;
```

Note that, as usual, OPL checks whether a data item is initialized exactly once in the model and raises an error otherwise. `initialize` instructions can have several parts, as in

```
{int} a[0..9];
{int} odd = {1, 3, 5, 7, 9};
{int} even = {2, 4, 6, 8, 10};

initialize {
```

⟨*Initialize*⟩ → Initialize ⟨*InitBody*⟩
⟨*InitBody*⟩ → ⟨*Expr*⟩ = ⟨*Expr*⟩
→ forall (⟨*ListParameter*⟩) ⟨*InitBody*⟩
→ if ⟨*Relation*⟩ then ⟨*InitBody*⟩ endif
→ if ⟨*Relation*⟩ then ⟨*InitBody*⟩ else ⟨*InitBody*⟩ endif
→ { ⟨*InitBody*⟩ }
→

Figure 4.10: The Syntax of Computed Initializations.

```
    forall(i in 0..5)
       a[i] = odd;
    forall(i in 6..9)
       a[i] = even;
};
```

Note that, once again, `initialize` instructions check whether each data item is initialized exactly once. The excerpt

```
{int} a[0..9];
{int} odd = {1, 3, 5, 7, 9};
{int} even = {2, 4, 6, 8, 10};

initialize {
    forall(i in 0..5)
       a[i] = odd;
    forall(i in 5..9)
       a[i] = even;
};
```

generates a runtime error, since `a[5]` is initialized twice. Figure 4.10 summarizes the syntax of computed initializations.

4.7.4 File Initializations

The last type of initialization involves listing the values of a data item in a file. For instance, the declaration

4. Data Modeling

```
struct Prec {
    int id;
    Task before;
    Task after;
};
{Prec} precedences < "prec.dat";
```

declares a set of records **precedences** whose initialization is given in the file **prec.dat**. The file **prec.dat** simply lists without punctuation all the values necessary to initialize **precedences**. For instance, the file **prec.dat** may contain the values

1 a1 a2 2 a2 a3 3 a3 a4 4 a4 a5

Note that the data need not be given on a single line. The file **prec.dat** may be organized as

1 a1 a2
2 a2 a3
3 a3 a4
4 a4 a5

Of course, OPL makes sure that the types of the values correspond to their declarations and that enough values are provided. Arrays can be initialized in the same fashion, e.g.

```
int array [1..10] < "arr.dat";
```

The file **arr.dat** may contain, for instance, the values

3 4 5 6 7 10 11 12 13 14

The only case in which punctuation symbols are needed in file initializations is when a set type is used inside another data structure. For instance, given the declaration

```
{int} a[1..3] < "a.dat";
```

the file **a.dat** may contain, for instance,

{3 4 5 6 7}
{3 4 6 7 8}
{2 3 4 5 9}

The braces are needed here to determine when the description of the set is completed. It is useful to stress that the file initialization is more memory-efficient than online or offline initialization and should be used for very large files.

5 Expressions and Constraints

This chapter describes expressions, relations, and constraints in OPL. Before proceeding further, it is important to recognize that expressions and relations are used in two fundamentally different ways in OPL:

1. to specify and initialize data;
2. to state constraints over variables.

In the first case, the expressions and relations do not contain variables or activities, since these variables have no value at this stage of the computation. These expressions and relations are said to be *ground* and they are subject to almost no restrictions. In the second case, of course, the expressions and relations may contain variables (and/or activities). Relations containing variables and activities are called *constraints* and are subject to a number of restrictions (e.g., float constraints must be linear or piecewise-linear).

The rest of this chapter is organized as follows. Section 5.1 describes expressions and relations in a general way, i.e., without concern for the restrictions imposed on constraints. Section 5.2 presents the general constraints allowed in OPL. Section 5.3 discusses the tools offered in OPL to state constraints.

5.1 Expressions and Relations

This section reviews how expressions and relations are built. Restrictions imposed on constraints are discussed in Section 5.2.

5.1.1 Data and Variable Identifiers

Since data and variable identifiers are the basic components of expressions, it is useful to review briefly how they are used to build expressions. If r is a record with a field capacity of type T, then r.capacity is an expression of type T. If a is a n-dimensional array of type T, a[$e_1, \ldots, e_n$] is an expression of type T, provided that e_i are well-typed. For instance, the excerpt

```
int limit[routes] = ...;
var int transp[r in routes] in 0..limit[r];
```

contains an expression limit[r] of type integer. Indices of arrays can be complex expressions. For instance, the excerpt

```
int nbFlights = ...;
range Flight 1..nbFlights;
enum Employee ...:
```

```
var int crew[Flight , Employee] in 0..1;
solve {
    ...
    forall(e in Employee)
        forall(i in 1..nbFlights - 2)
            crew[i,e] + crew[i+1,e] + crew[i+2,e] >= 1;
};
```

contains an integer expression `crew[i+1,e]` whose first index is itself an integer expression. The excerpt

```
forall(w in Women)
    forall(i in R)
        rankWomen[w,favoriteMen[w,i]] = i;
```

contains an expression `favoriteMen[w,i]` as the second index of array `rankWomen`. Indices of arrays can also contain variables. For instance, the excerpt

```
range Warehouses ...;
range Customers ...;
var Warehouses supplier[Customers];
var int open[Customers] in 0..1;
solve {
    ...
    forall(c in Customers)
        open[supplier[c]] = 1;
}
```

contains an expression `Open[supplier[c]]` and `supplier[c]` is an integer variable. The ability to use variables in indices makes possible some very compact models for some combinatorial optimization problems.

5.1.2 Integer Expressions

Integer expressions are constructed from integer constants, integer data, integer variables, and the traditional integer operators such as +, -, *, /, mod; the operator / represents integer division (e.g., $8/3 = 2$) and the operator mod represents integer remainder. OPL also supports the operator abs, which returns the absolute value of its argument, and the constant maxint, which represents the largest integer representable in OPL. Users should, of course, be aware that expressions involving large integers may produce overflow.

5.1.3 Float Expressions

Float expressions are constructed from floats, float data and variables, as well as operations such as +, -, /, *. In addition, OPL contains a float constant `infinity` to represent ∞ and a variety of operations such as `abs, sqrt, ceil, distToInt, floor, frac, nearest`, and `trunc`. Given a float a, `abs(a)` returns the absolute value of a, `sqrt(a)` the square root of a, `ceil(a)` the smallest integer greater than or equal to a, `distToInt(a)` the distance from a to the nearest integer, `floor(a)` the largest integer smaller or equal to a, `frac(a)` the fractional part of a, `nearest(a)` the nearest integer to a (it returns `ceil(a)` when `frac(a)=0.5`), and `trunc(a)` the integer part of a. Integers and floats can be mixed inside the same expressions: integers are automatically converted to type `float` in such expressions. Note that the results of all these float functions are of type `float`. The function `ftoi` (float to integer) can be used to convert an integer represented as float into an integer represented as an integer. For instance, the declaration

```
int i = ftoi(nearest(4.7));
```

is valid, while the declaration

```
int i = nearest(4.7);
```

produces a semantic error. The result of function `ftoi` is undefined when the integer cannot be represented exactly in type `int`.

5.1.4 Enumerated Expressions

Enumerated expressions are constructed from enumerated values, enumerated data and variables, as well as a number of functions over enumerated types and values. OPL supports various functions over enumerated types. The function `first` and `last`, when applied to an enumerated type T, returns the first and the last enumerated values of T. For instance, the excerpt

```
enum Days { Monday, Tuesday, Wednesday, Thursday, Friday, Saturday, Sunday };
Days f = first(Days);
Days l = last(Days);
```

initializes f and l to `Monday` and `Sunday` respectively. This also indicates that enumerated sets (and, in fact, all sets) in OPL are totally ordered according to their order of appearance in the initialization. The function `card`, when applied to an enumerated type, returns the number of enumerated values in the type. For instance, `card(Days)` returns 7.

A number of functions are also available on enumerated values. Function `ord(v)` returns the position of value v in the enumerated type, with the understanding that positions start at zero (as in C and PASCAL). For instance, `ord(Tuesday) = 1` and `ord(Sunday) = 6`. The function `next(v)` returns the successor to v if it exists and raises an error otherwise (e.g., `next(Monday) = Tuesday`) and function `prev(v)` returns the predecessor to `v` if it exists and raises an error otherwise (e.g., `prev(Tuesday) = Monday`). These functions can be generalized to give the ith successor (resp. predecessor) of a value ($i \geq 0$). For instance, `next(Monday,3) = Thursday`. Once again, this value should be well-defined or an execution error is produced. OPL also supports circular versions of these functions (i.e., `nextc` and `prevc`) which consider that the successor of the last element of the type is the first element and vice-versa for the predecessor. For instance, `nextc(Sunday) = Monday`, while `next(Sunday)` raises an error.

Enumerated expressions can be used as indices or indexed by other enumerated expressions. For instance, the excerpt

```
enum Men ...;
enum Women ...;
var Women wife[Men];
var Men husband[Women];
...
solve {
    ...
    forall(m in Men)
        husband[wife[m]] = m;
    ...
};
```

is an enumerated expression `husband[wife[m]]` where the enumerated array `husband` is indexed by another enumerated expression `wife[m]`. Note that, in this example, `husband` and `wife` are both arrays of variables.

5.1.5 Aggregate Operators

Integer and float expressions can also be constructed using aggregate operators for computing summations (`sum`), products (`prod`), minima (`min`), and maxima (`max`) of a collection of related expressions. For instance, the excerpt

```
int capacity[Routes] = ...;
int minCap = min(r in Routes) capacity[r];
```

uses the aggregate operator `min` to compute the minimum value in array `capacity`. The form of the formal parameters in these aggregate operators is very general and is discussed at length in Chapter 6.

5. Expressions and Constraints

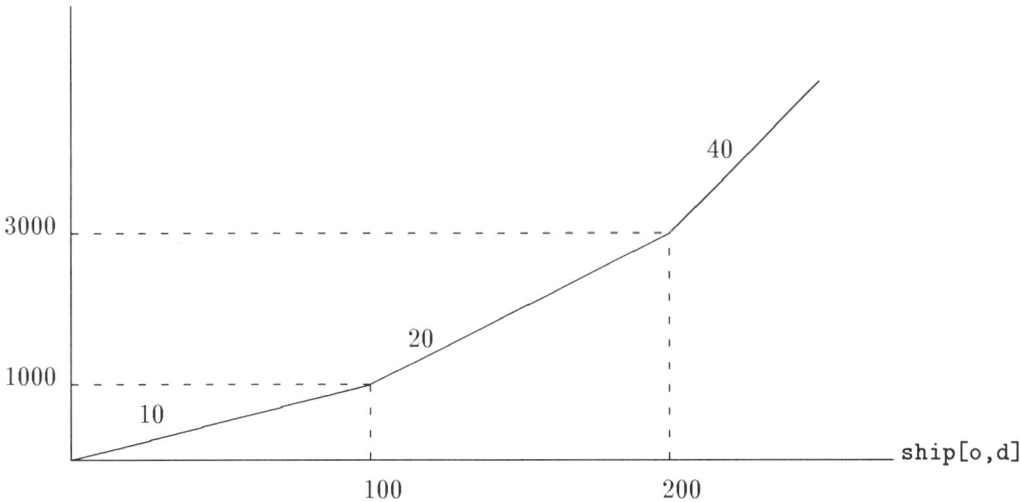

Figure 5.1: A Piecewise-Linear Function.

5.1.6 Piecewise Linear Functions

OPL also supports piecewise-linear functions, which are important in many applications. Piecewise-linear functions are discussed at length in Chapter 9. Piecewise-linear functions are often specified by giving a set of slopes, a set of breakpoints at which the slopes change, and the value of the functions at a given point. Consider, for instance, a transportation problem in which the transportation cost ship[o,d] between two locations o and d depends on the size of the shipment. The piecewise-linear expression

`piecewise{10 -> 100;20 -> 200,40}(0,0) ship[o,d];`

describes the piecewise-linear function of ship[o,d] depicted in Figure 5.1. The function has slopes 10, 20, and 40, breakpoints 100 and 200, and evaluates to 0 at point 0. In other words, the piecewise-linear expression is equivalent to the expression

`10 * ship[o,d]`

when ship[o,d] <= 100, to

`10 * 100 + 20 * (ship[o,d] - 100)`

when 100 <= ship[o,d] <= 200 and to

`10 * 100 + 20 * 100 + 40 * (ship[o,d] - 200)`

otherwise. By default, OPL assumes that a piecewise-linear function evaluates to zero at the origin, so that the above piecewise-linear function could actually be written as

```
piecewise{10 -> 100;20 -> 200,40} ship[o,d];
```

The above piecewise-linear function has a fixed number of pieces, but OPL also allows generic pieces. The number of pieces may then depend on the input data, as in

```
piecewise {
    forall(i in 1..n) slope[i] -> breakpoint[i];
    slope[n+1];
} ship[o,d];
```

This piecewise-linear function is equivalent to

```
slope[1] * ship[o,d]
```

when `ship[o,d] <= breakpoint[1]`, to

slope[1] * breakpoint[1] +
$\sum_{i=2}^{k-1}$ slope[i] * (breakpoint[i] - breakpoint[i-1]) +
slope[k] * (ship[o,d] - breakpoint[k-1])

when `breakpoint[k-1] < ship[o,d] <= breakpoint[k]` ($1 \leq k \leq n$), and to

slope[1] * breakpoint[1] +
$\sum_{i=2}^{n}$ slope[i] * (breakpoint[i] - breakpoint[i-1]) +
slope[k] * (ship[o,d] - breakpoint[n])

otherwise. Note that there may be several generic pieces in piecewise-linear functions. It is important to stress that breakpoints and slopes in piecewise-linear functions must always be ground and that the breakpoints must be strictly increasing. Chapter 9 discusses piecewise-linear functions in more detail.

5.1.7 Set Expressions

Set data can be initialized by set expressions, as mentioned in Chapter 4. These expressions are constructed from previously defined sets and the set operations `union`, `inter`, `diff`, and `symdiff`. For instance,

```
{int} s1 = {1,2,3};
{int} s2 = {1,4,5};
{int} i = s1 inter s2;
```

```
{int} u = s1 union s2;
{int} d = s1 diff s2;
{int} sd = s1 symdiff s2;
```

initializes i to {1}, u to {1,2,3,4,5}, d to {2,3}, and sd to {2,3,4,5}. Note that explicitly defined sets (e.g., {2,3,4,5} cannot appear in these expressions. Set expressions can also be constructed from some aggregate operators. For instance, the excerpt

```
{int} a[R] = ...;
{int} U = union(s in R) a[s];
{int} i = inter(s in R) a[s];
```

illustrates how to compute the union and intersection of a collection of sets. In addition, set expressions can be constructed from ranges. For instance, the excerpt

```
{int} s = 1..10;
{int} ps = 1..10 by 2;
```

initializes s to the finite set $\{1, 2, ..., 10\}$ and ps to the finite set $\{1, 3, 5, 7, 9\}$. It is important to stress that the range 1...10 takes constant space, while the set s takes space proportional to the number of elements in the range.

5.1.8 Relations in Expressions

Expressions can also be constructed in terms of relations because OPL views every relation as a 0-1 integer. For instance, the excerpt

```
{int} s[0..n] = ...;
{int} occur[i in 0..n] = sum(j in 0..n) (s[j] = i);
```

initializes occur[i] with the number of times i occurs in array s. This initialization uses an aggregate operator over relations. A relation s[j] = i evaluates to one when true and to zero otherwise. Traditional Boolean connectives can be used to combine these relations and the resulting relation can also be used inside expressions. For instance, the excerpt

```
sum(s in Squares) (x[s] <= p & p <= x[s] + size[s]) = n;
```

takes the conjunction of two relations. It could actually be rewritten more compactly as

```
sum(s in Squares) (x[s] <= p <= x[s] + size[s]) = n;
```

Other connectives such as \/ (disjunction), not (negation), => (implication), and <=> (equivalence) can also be used.

5.1.9 Reflective Functions

Reflective functions are used to obtain information about the current state of the computation, e.g., the domains of the variables or the current value of a floating-point variable in the linear relaxation. Their arguments are in general variables. Table 5.1 gives their signatures, preconditions, and semantics. To express the preconditions, we use $\underline{e}$ to indicate that the expression e must be ground and, in the semantics, we use `dom(c)` to represent the domain of c in the current computation state. `UnaryResource` and `DiscreteResource` can also be abbreviated by `Unary` and `Discrete` for compactness. Note that a signature of the form `dmin(int c)` specifies that `dmin` expects an argument of type integer but the argument may be a data or a variable (i.e., the signature constrains only the type). OPL also makes it is possible to use `v.rc` as an abbreviation of `reducedCost(v)`.

Reflective functions should only be used to express heuristics and search procedures, since they are operational in nature and rely on the state of the variables at some execution step. As mentioned previously, one of the appealing features of constraint programming in general, and OPL in particular, is its declarative nature: statements can be given a meaning without relying on knowledge of how they are executed. However, the use of reflective functions outside the specification of heuristics jeopardizes this benefit.

5.1.10 Relations

Like expressions, relations can have various types in OPL. Integer relations are constructed from integer expressions and the traditional relational operators =, <> (not equal), >=, >, <, and <=. Float relations are constructed from float expressions and the same relational operators. Enumerated relations are constructed from enumerated expressions and support the same operators as well. Of course, only data from the same enumerated types can be compared. Relations can also be used in sequence, as shown before. For instance,

```
low <= x <= up;
```

can be used instead of

```
low <= x; x <= up;
```

Similarly, a relation

```
a <= b <= c <= d;
```

can be used instead of

```
a <= b; b <= c; c <= d;
```

Note that

5. Expressions and Constraints

Function	Type	Semantics
dmin(int c)	int	the minimum value in dom(c)
dmax(int c)	int	the maximum value in dom(c)
dmin(float c)	float	the minimum value in dom(c)
dmax(float c)	float	the maximum value in dom(c)
dsize(int c)	int	the size of dom(c)
dsize(enum c)	int	the size of dom(c)
dnexthigher(int c, int e)	int	the smallest value greater than e in dom(c) or e if none exists
dnextlower(int c, int e)	int	the largest value smaller than e in dom(c) or e if none exists
regretdmin(int c)	int	dnexthigher(c,dmin(c)) - dmin(c)
regretdmax(int c)	int	dmax(c) - dnextlower(c,dmax(c))
simplexValue(int c)	float	the value of c in the simplex solution
simplexValue(float c)	float	the value of c in the simplex solution
reducedCost(int c)	float	the reduced cost of c in the simplex solution
reducedCost(float c)	float	the reduced cost of c in the simplex solution
bound(int c)	int	1 if dmin(c) = dmax(c) and 0 otherwise
bound(enum c)	int	1 if dsize(c) = 1 and 0 otherwise
bound(float c)	int	1 if dmin(c) = dmax(c) and 0 otherwise
nbOccur(int e,[int] id)	int	the number of occurrences of e in array id
isRanked(Unary)	int	1 if the resource is ranked and 0 otherwise
isRanked([Unary])	int	1 if all resources are ranked and 0 otherwise
nbPossibleFirst(Unary)	int	number of activities that can be ranked first
nbPossibleLast(Unary)	int	number of activities that can be ranked last
isPossibleFirst(Unary r,Activity a)	int	true iff a can potentially be ranked first on r
isPossibleLast(Unary,Activity)	int	true iff a can potentially be ranked last on r
localSlack(Unary)	int	local slack of the resource
globalSlack(Unary)	int	global slack of the resource
localSlack(Discrete)	int	local slack of the resource
globalSlack(Discrete)	int	global slack of the resource

Table 5.1: Reflective Functions in OPL.

Class	Operator	Precedence
Logical	<=>	0
	=>	1
	\/	2
	&	3
	not	4
Relational	=, <=, >=, <, >, <>	5
	in, not in, subset, requires, consumes	5
	provides, produces, precedes	5
Binary	+, -, union, diff, symdiff	6
Unary	+, -	7
Aggregate	sum, max, min	7
	union	7
Binary	*, /, mod, inter	8
Aggregate	prod, inter	9

Table 5.2: Operator Precedences.

```
a <= b <= c;
```

is fundamentally different from

```
(a <= b) <= c;
```

This last constraint is a higher-order inequality between (a <= b) and c. Finally, note that a relation (a <= b <= c) has the same truth value as a <= b & b <= c.

OPL also supports the relations e in S and e notin S, where e has type T and S has type "set of T", to check membership and non-membership in a set. Finally, as mentioned previously, relations can also be combined by the traditional logical connectives: negation (not), disjunction \/, conjunction &, implication =>, and equivalence <=>.

5.1.11 Syntax

Figures 5.2 and 5.3 describe the syntax of expressions and relations in OPL. In addition, Table 5.2 specifies the precedence of the operators. OPL uses the conventions that higher numbers have precedence over lower ones, that expressions containing operators at the same level are parsed from left to right and top to bottom, and that parentheses can overwrite the order of evaluation.

5. Expressions and Constraints

$\langle Expr \rangle$	$\to$	+ $\langle Expr \rangle$
	$\to$	- $\langle Expr \rangle$
	$\to$	$\langle Expr \rangle$ $\langle Operator \rangle$ $\langle Expr \rangle$
	$\to$	$\langle AggregateOperator \rangle$ ($\langle ListParameter \rangle$) $\langle Expr \rangle$
	$\to$	`piecewise` { $\langle ListPiece \rangle$ } $\langle Composite \rangle$
	$\to$	`piecewise` { $\langle ListPiece \rangle$ } ($\langle Expr \rangle$, $\langle Expr \rangle$) $\langle Composite \rangle$
	$\to$	($\langle Relation \rangle$)
	$\to$	($\langle Expr \rangle$)
	$\to$	$\langle Composite \rangle$
	$\to$	`Integer`
	$\to$	`Float`
	$\to$	`infinity`
	$\to$	`maxint`
$\langle ListPiece \rangle$	$\to$	$\langle Expr \rangle$
	$\to$	$\langle Expr \rangle$ -> $\langle Expr \rangle$; $\langle ListPiece \rangle$
	$\to$	`forall` ($\langle ListParameter \rangle$) $\langle Expr \rangle$ -> $\langle Expr \rangle$; $\langle ListPiece \rangle$
$\langle Composite \rangle$	$\to$	$\langle Id \rangle$ $\langle Dereference \rangle$
$\langle Dereference \rangle$	$\to$	$\underline{\{ \langle Deref \rangle \}}$
$\langle Deref \rangle$	$\to$	$\overline{[\langle Actuals \rangle]}$
	$\to$	. $\langle Id \rangle$
	$\to$	($\langle Actuals \rangle$)
$\langle Actuals \rangle$	$\to$	$\langle EltInit \rangle^+$
$\langle Operator \rangle$	$\to$	+ | - | * | / | mod | union | diff | inter | symdiff
$\langle AggregateOperator \rangle$	$\to$	sum | min | max | union | prod | inter

Figure 5.2: The Syntax of Expressions.

$\langle Relation \rangle$ → $\langle Expr \rangle\ \langle RelSfx \rangle^+$
→ $\langle Composite \rangle\ \langle SchedOp \rangle\ \langle Composite \rangle$
→ $\langle Composite \rangle\ \langle SchedOp \rangle\ (\langle Expr \rangle)\ \langle Composite \rangle$
→ `not` $\langle Relation \rangle$
→ $\langle Relation \rangle\ \langle LogicalOp \rangle\ \langle Relation \rangle$
→ $\langle Expr \rangle$ `in` $\langle Expr \rangle$
→ $\langle Expr \rangle$ `not in` $\langle Expr \rangle$
→ $\langle Expr \rangle$ `subset` $\langle Expr \rangle$
→ $\langle Composite \rangle$ `precedes` $\langle Composite \rangle$
→ `activityHasSelectedResources(`$\langle Composite \rangle$`,`$\langle Composite \rangle$`,`$\langle Composite \rangle$`)`

$\langle RelSfx \rangle$ → $\langle RelOp \rangle\ \langle Expr \rangle$

$\langle RelOp \rangle$ → `=` | `>=` | `<=` | `>` | `<` | `<>`
$\langle SchedOp \rangle$ → `requires` | `consumes` | `provides` | `produces`
$\langle LogicalOp \rangle$ → `&` | `∨` | `=>` | `<=>`

Figure 5.3: The Syntax of Relations.

$\langle Cstr \rangle$
→ $\langle Relation \rangle$
→ $\langle Composite \rangle$: $\langle Relation \rangle$
→ forall ($\langle ListParameter \rangle$) $\langle Cstr \rangle$
→ if $\langle Relation \rangle$ then $\langle Cstr \rangle$ endif
→ if $\langle Relation \rangle$ then $\langle Cstr \rangle$ else $\langle Cstr \rangle$ endif
→ { $\langle Cstr \rangle^*$ }
→ circuit ($\langle Composite \rangle$)
→ alldifferent ($\langle CstrComposite \rangle$) $\langle Propagation \rangle$
→ distribute ($\langle Composite \rangle$, $\langle Composite \rangle$, $\langle CstrComposite \rangle$)
→ capacityMax ($\langle Composite \rangle$, $\langle Expr \rangle$, $\langle Expr \rangle$, $\langle Expr \rangle$)
→ capacityMin ($\langle Composite \rangle$, $\langle Expr \rangle$, $\langle Expr \rangle$, $\langle Expr \rangle$)
→ periodicBreak ($\langle Composite \rangle$, $\langle Expr \rangle$, $\langle Expr \rangle$, $\langle Expr \rangle$)
→ break ($\langle Composite \rangle$, $\langle Expr \rangle$, $\langle Expr \rangle$)
→ breakOnDuration ($\langle Composite \rangle$, $\langle Expr \rangle$, $\langle Expr \rangle$)
→

$\langle Propagation \rangle$ → onValue | onRange | onDomain |

$\langle CstrComposite \rangle$ → $\langle Composite \rangle$
→ all ($\langle ListParameter \rangle$) $\langle Expr \rangle$

Figure 5.4: The Syntax of Constraints.

5.2 Constraints

Constraints are a subset of relations. This section specifies the constraints supported by OPL and discusses various subclasses of constraints to illustrate the rich support available for modeling combinatorial optimization applications. Figure 5.4 summarizes the syntax of constraints in OPL.

5.2.1 Float Constraints

Float constraints in OPL are restricted to be linear or piecewise linear. OPL has efficient algorithms for solving linear real constraints, but in general these algorithms do not apply to nonlinear problems. Note that the linearity requirement precludes the use of relations with variables in constraints and the use of non-ground expressions as indices of float arrays. In addition, operators <>, <, and > are not allowed for float constraints. Note, however, that integers and integer variables may occur in a float constraint, provided that the constraint remains linear.

5.2.2 Discrete Constraints

Discrete constraints are arbitrary integer or enumerated relations, possibly containing variables. These constraints must be well-typed, but no restrictions are imposed on them. It is, however, useful to review subclasses of these constraints to illustrate the functionalities of OPL.

Basic Constraints Basic discrete constraints are constructed from discrete data, discrete variables, and the arithmetic operators and functions defined previously. For instance, the excerpt

```
var int freq[Freqs] in 0..256;
solve {
    forall(f,g in Freqs)
        abs(freq[f] - freq[g]) > 16;
...
};
```

generates distance constraints between integer variables. OPL uses discrete constraints to reduce the domain of discrete variables by applying various local consistency algorithms.

Logical Combinations of Constraints Discrete constraints can also be combined with traditional logical connectives. For instance, the excerpt

```
enum Tasks ...
struct Disj {
    Tasks before;
    Tasks after;
};
{Disj} disjunctions = ...;
int duration[Tasks] = ...;
var int start[Tasks] in 0..10000;
solve {
    forall(<b,a> in disjunctions)
        start[b] >= start[a] + duration[a] \/ start[a] >= start[b] + duration[b];
};
```

states disjunctions of basic constraints. The constraint

```
start[b] >= start[a] + duration[a] \/ start[a] >= start[b] + duration[b]
```

has the obvious declarative meaning: task b must be scheduled after task a or vice-versa. Operationally, OPL makes sure that at least one of the disjuncts is consistent with the constraint store. In particular, if one of the disjuncts is not consistent with the constraint store, the other disjunct is added to the store.

Higher-order Constraints Discrete constraints can also include other discrete constraints as a part of their expressions. Once again, an embedded constraint is associated with 0-1 integer variable that becomes 1 when the constraint can be proven true, 0 when the constraint can be proven false, and remains undetermined otherwise. For instance, in the magic series statement,

```
int n << "Number of Variables: ";
range Range 0..n-1;
range Domain 0..n;
var Domain s[Range];
solve {
   forall(i in Range)
      s[i] = sum(j in Range) (s[j] = i);
};
```

the constraint

```
s[i] = sum(j in Range) (s[j] = i);
```

is a higher-order constraint associating a 0-1 variable with each constraint s[j] = i. Whenever OPL can show that s[j] = i, the 0-1 variable becomes true (i.e., is given the value 1). Whenever OPL can prove that s[j] <> i, the 0-1 variable becomes false (i.e., is given the value 0). Higher-order constraints are useful for expressing cardinality constraints. For instance, the constraint

```
sum(j in S) (a[j] = 2) >= 3
```

says that the array a must contain at least three occurrences of the value 2.

Variables as Indices As mentioned previously, OPL makes it possible to use variables to index arrays. For instance, in the stable marriage statement

```
enum Women ...;
enum Men ...;
int rankW[Women,Men] = ...
int rankM[Men,Women] = ...
```

```
var Women wife[Men];
var Men husband[Women];
solve {
   forall(m in Men)
      husband[wife[m]] = m;
   forall(w in Women)
      wife[husband[w]] = w;
   forall(m in Men & o in Women)
      rankM[m,o] < rankM[m,wife[m]] => rankW[o,husband[o]] < rankW[o,m];
   forall(w in Women & o in Men)
      rankW[w,o] < rankW[w,husband[w]] => rankM[o,wife[o]] < rankM[o,w];
}
```

the implication

`rankM[m,o] < rankM[m,wife[m]] => rankW[o,husband[o]] < rankW[o,m]`

indexes the array `rankM` using a variable `wife[m]` as second index. The expression

`rankM[m,wife[m]]`

should be viewed as a variable whose domain is the set of possible values in the array `rankM`, which can be obtained by assigning `wife[m]` to the values in its domain. Variables can also be used to index arrays of variables. For instance, the constraint

`wife[husband[w]] = w;`

indexes the array of variables `wife` with the variable `husband[w]`. Once again, the expression `wife[husband[w]]` is best thought of as a variable whose set of values are all the values that can be obtained by assigning `husband[w]` and the variables in array `wife` in all possible ways. Of course, OPL uses constraints to reduce the domain of `husband[w]` and thus the domains of all variables in `wife`.

It is important to stress that the current implementation of OPL is more efficient when variables are used to index the last dimension of a multi-dimensional array.

Multi-directionality of Combinations of Constraints It is important to stress that constraints are multi-directional and to highlight an important consequence. When constraints are combined with logical connectives, OPL states all constraints in the expression and uses their truth values with respect to the constraint store to deduce whether other constraints must be true or false. For instance, an implication of the form `a => b`, where `a` and `b` are constraints, propagates two types of information:

1. when a is entailed by the constraint store, then b is added to the constraint store;
2. when the negation of b is entailed by the constraint store, then the negation of a is added to the constraint store.

Note that both a and b are used by OPL. As a consequence, they must be well-defined. A typical mistake in this context is to assume that b is only used when a is entailed by the store, as in the excerpt

```
var int m[1..10] in 0..10;
solve {
   forall(i in 1..10)
      m[i] <> 0 => m[m[i]] = i;
};
```

The problem here is that if, say, m[2] = 0, the constraint m[0] = i is evaluated by OPL. But this constraint is not well-defined and OPL will fail. This result may seem counter-intuitive but it is a direct consequence of the multi-directionality of constraints. Use the data-driven constructs described in Section 7.11 as an alternative to implication when a constraint must be guarded by a condition.

Global Constraints OPL also offers a variety of global constraints over discrete values. The constraint **alldifferent** expects an array of discrete variables/values and holds if all elements of the array are given a different value. The pruning achieved by this constraint can be specified by using the keywords onValue, onRange, onDomain. When onValue is used, OPL guarantees that, at any computation point, the variables in the array do not have the values of the already assigned variables inside their domain. When onDomain is used, OPL guarantees that, for each value in the domain of any given variable, there exist values in the domains of the remaining variables such that the constraint is satisfied. When onRange is used, the constraint enforces a pruning stronger than onValue and weaker than onDomain by reasoning only on the bounds. The constraint

alldifferent(a)

is in fact equivalent to

alldifferent(a) onValue

specifying the default of the system.

The constraint circuit expects an array of integer variables, say succ, whose index set is the range 1..n. Each value in the range 1..n corresponds to a node in a graph and the value succ[i] represents the successor of node i in the graph. As a consequence, all elements in succ must range in 1..n. The constraint circuit(succ) holds if the graph sodefined is a Hamiltonian circuit or, more precisely, that the sequence

$$(1, v_1), (v_1, v_2), \ldots, (v_{n-1}, v_n), (v_n, v_{n+1})$$

where

$$\begin{cases} v_1 = \mathtt{succ}[1] \\ v_i = \mathtt{succ}[v_{i-1}] \quad (2 \leq i \leq n+1) \end{cases}$$

is a Hamiltonian circuit. Operationally, constraint circuit removes values that would produce a subtour that is not a Hamiltonian circuit.

The constraint distribute captures a class of cardinality constraints. It expects three one-dimensional arrays card, value, and base; in addition, the arrays card and values must have the same index set. The constraint distribute(card,value,base) then holds if card[i] is the number of occurrences of value[i] in array base. If the index set of card and value is S and if the index set of base is R, then distribute(card,value,base) is equivalent to, but more efficient than

```
forall(i in S)
    card[i] = sum(j in R) (value[i] = base[j]);
```

Global Constraints on Aggregate Arrays It is sometimes convenient to state a global constraint on some elements on an array. Without additional language support, it would be necessary to declare an additional array of variables and to state equations between the new variables and the variables of interest. To remove this inconvenience, OPL provides an aggregate operator to select elements from an array. The statement

```
var int v[1..4,1..5] in 0..100;
solve {
    alldifferent(all(i in 1..4,j in 1..5:  i + j < 5) v[i,j]);
};
```

illustrates this functionality. It selects all the elements v[i,j] such that i+j<5 and states that they must be all different. The third argument of global constraint distribute can also use an aggregate array instead of a classical array.

5.2.3 Scheduling Constraints

OPL also provides some specific support for scheduling applications.

5.2.3.1 Precedence Constraints

Precedence constraints between activities can be expressed easily in terms of the starting date, the ending date, and the duration of the activities. For instance, a precedence constraint between two activities `a` and `b` can be specified as `b.start >= a.end` However, OPL has a specific constraint for expressing precedence constraints, which is recommended since it produces better visualizations of the results. For instance, a precedence constraint between two activities `a` and `b` can be specified as `a precedes b`.

5.2.3.2 Resource Constraints

The most fundamental constraints in scheduling applications are of course the resource constraints that link the activities and the resources. This section describes these constraints for the various kinds of resources.

Unary Resources Once unary resources are declared, it is possible to specify which activities require them. To specify that an activity `excavation` requires a resource `crane` in OPL, it is sufficient to write the constraint

```
excavation requires crane;
```

This constraint specifies that activity `excavation` requires unary resource `crane` during its execution, and OPL makes sure that no two activities requiring the same unary resource are scheduled at the same time. In addition, OPL also uses these constraints to update the starting and ending dates of the activities using an edge-finder algorithm.

OPL provides two constraints to specify the breaks of unary resources. In general, breaks are periodic (like weekends) and it is convenient to specify them using the constraint `periodicBreak`, as in the excerpt

```
UnaryResource aCrane;
...
solve {
    periodicBreak(aCrane,5,2,7);
    ...
};
```

which specifies that the unary resource `aCrane` has a break every seven days, that its first break is on day 5, and that the duration of the break is 2. The signature of `periodicBreak` is thus

```
periodicBreak(<UnaryResource>,<StartingDate>,<Duration>,<Periodicity>)
```

Individual breaks can also be specified, as in the excerpt

```
break(aCrane,10,12);
```

indicating that resource `aCrane` has a break over the interval [10,12[. The same break can be specified as

```
breakOnDuration(aCrane,10,2);
```

Discrete Resources Activities can require discrete resources in the same way as unary resources. In addition, an activity can require a certain amount of the discrete resource. For instance, the constraint

```
a requires(2) hammer;
```

specifies that activity `a` requires two hammers. The requested amount can be an arbitrary integer expression, possibly containing variables. The underlying algorithms in OPL make sure that at no times the requested amount exceed the capacity of the resource. In fact, three levels of pruning can be achieved in OPL for discrete resources: the default level, the *disjunctive* level, and the *edge-finder* level. The *disjunctive* level can be requested by declarations of the form

```
DiscreteResource hammer(3) using disjunctive;
```

This level of pruning ensures that, if the total demand of a set of activities requires more than the capacity of the resource, then at least one of these activities is scheduled before or after another activity in the set. The *edge-finder* level can be requested by declarations of the form

```
DiscreteResource hammer(3) using edgeFinder;
```

The *edgeFinder* level generalizes the edge-finder algorithm of unary resources to discrete resources. It tries to deduce which activities must be scheduled first (or last) in a set whose total demand exceeds the capacity of the resource. The level of propagation appropriate depends of course on the application at hand. As mentioned previously, a constraint

```
a requires(2) r;
```

specifies that activity `a` requires the resource `r` during its execution. As a consequence, as soon as activity `a` terminates, its requested capacity is returned to the resource `r` and is available for other activities. For some applications (e.g., when the discrete resource denotes a budget), the requested capacity should not be returned to the resource: it is consumed forever. This functionality is obtained in OPL by constraints of the form

```
a consumes(2) r;
```

Note that it is also possible to use `consumes` for unary resources although, in general, this is not particularly useful.

The capacity of a discrete resource may also vary in time, generalizing the concept of breaks in unary resources. The capacity of a discrete resource over time can be specified with constraints of the form

```
capacityMax(<DiscreteResource>,<Start>,<End>,<Cap>)
capacityMin(<DiscreteResource>,<Start>,<End>,<Cap>)
```

A constraint `capacityMax(d,s,e,c)` specifies that the capacity of discrete resource `d` required by activities over the interval `[s,e)` is at most `c`, while a constraint `capacityMin(d,s,e,c)` specifies that the capacity of discrete resource `d` required by activities over the interval `[s,e)` is at least `c`. Note that this last constraint in fact constrains the scheduling of activities to make sure that they satisfy this capacity constraint.

Reservoirs Reservoirs support the same constraints as discrete resources. In addition, they also support constraints allowing activities to provide and produce some units of the resources. Assuming that `plumbing` and `tank` are declared as reservoirs,

```
a requires(2) plumbing;
```

specifies that activity `a` requires two units of plumbing during its execution. Activities can also consume the reservoir, as in

```
a consumes(2) tank;
```

which specifies that activity `a` consumes 2 units of reservoir `tank` (and does not return these two units at the end of its execution). Constraint

```
b provides(2) plumbing;
```

specifies that activity `b` provides two units of plumbing during its execution, while constraint

```
b produces(2) oil;
```

specifies that activity `b` produces two units of plumbing from its ending date to the horizon. Of course, providing and producing are the counterparts to requiring and consuming.

Alternative Resources Alternative resources can be used in `requires` constraints as unary resources. For instance, assuming the declarations

```
UnaryResource worker[1..10];
AlternativeResources s(worker);
Activity a;
```

the constraint

```
a requires s;
```

specifies that activity `a` requires one of the unary resources in `s`. In addition to the `requires` constraints, OPL also provides an interesting tool for manipulating alternative resources: the constraint `activityHasSelectedResource(a,s,u)`, which holds if activity `a` has selected resource `u` in alternative resource `s`. Of course, this constraint can be negated and used as a higher-order constraint to express a variety of concepts.

5.3 Stating Constraints

Figure 5.5 describes the syntax of instructions in OPL. Instructions are responsible for stating the problem constraints and, in optimization problems, the objective function. Constraints are stated in OPL either by a `solve` instruction, as in

```
solve {
    forall(i in 0..8)
        s[i] = sum(j in 0..8) (s[j] = i);
};
```

or by an optimization instruction, as in

```
minimize
    sum(p in Products) (insideCost[p]*inside[p] + outsideCost[p]*outside[p])
subject to {
    forall(r in Resources)
        sum(p in Products) consumption[p,r] * inside[i] <= capacity[r];
    forall(p in Products)
        inside[p] + outside[p] >= demand[p];
};
```

Note that the optimization instructions require an objective function of type integer or float.

5.3.1 Linear Relaxation

The keyword `with linear relaxation` may be added after the `solve`, `minimize`, or `maximize` instructions. Its effect is to add the linear relaxation of any integer constraint to the constraint

5. Expressions and Constraints

$\langle Instr \rangle \;\to\;$ `solve` $\langle Relax \rangle \; \langle Cstr \rangle$;
$ \;\to\;$ `minimize` $\langle Relax \rangle \; \langle Expr \rangle$ `subject to` $\langle Cstr \rangle$;
$ \;\to\;$ `maximize` $\langle Relax \rangle \; \langle Expr \rangle$ `subject to` $\langle Cstr \rangle$;

Figure 5.5: The Syntax of Instructions.

---

store. In other words, the constraint store receives the relaxation of an integer constraint c whenever c becomes a float constraint when all its integer variables are considered of type float. This feature is discussed again in Chapter 10.

5.3.2 Universal Quantifiers

OPL provides a universal quantifier to declare a collection of closely related constraints. For instance, the excerpt

```
forall(i in 0..8)
    s[i] = sum(j in 0..8) (s[j] = i);
```

states the constraints

```
s[0] = sum(j in 0..8) (s[j] = 0);
...
s[8] = sum(j in 0..8) (s[j] = 8);
```

It is possible to use nested `forall` statements, as in

```
forall(i in 1..8)
    forall(j in [i+1..8])
        queen[i] <> queen[j]
```

`forall` statements can also contain a block, e.g.,

```
forall(i in 1..8)
    forall(j in [i+1..8]) {
        queen[i] <> queen[j];
        queen[i] - i <> queen[j] - j;
        queen[i] + i <> queen[j] + j;
    };
```

Of course, the above statement can be written more concisely as

```
forall(ordered i,j in 1..8) {
    queen[i] <> queen[j];
    queen[i] - i <> queen[j] - j;
    queen[i] + i <> queen[j] + j;
};
```

OPL offers a rich set of formal parameters, as described in more detail in Chapter 6.

5.3.3 Conditional Statements

`if-then-else` statements make it possible to state constraints conditionally, as in

```
forall(<f,g,d> in Distances)
    if d > 1 then
        abs(freq[f] - freq[g]) >= d
    else
        freq[f] <> freq[g]
    endif;
```

Conditions in `if-then-else` statements must be ground, i.e., they must not contain variables (except, possibly, in reflective functions). Implications of constraints can be used instead when conditions contain variables.

5.3.4 Naming Constraints

Constraints can also be given a name, as in

```
constraint capacityCons[Resources];
constraint demandCons[Products];

minimize
    sum(p in Products) (insideCost[p]*inside[p] + outsideCost[p]*outside[p])
subject to {
    forall(r in Resources)
        capacityCons[r]: sum(p in Products) consumption[p,r] * inside[i] <= capacity[r];
    forall(p in Products)
        demandCons[p]: inside[p] + outside[p] >= demand[p];
};
```

This statement is equivalent to the statement

```
minimize
    sum(p in Products) (insideCost[p]*inside[p] + outsideCost[p]*outside[p])
subject to {
    forall(r in Resources)
        sum(p in Products) consumption[p,r] * inside[i] <= capacity[r];
    forall(p in Products)
        inside[p] + outside[p] >= demand[p];
};
```

shown earlier in this section. The only difference is that the constraints have been given a name that can be used to display the data. This functionality is often useful to find which constraints are tight in applications.

6 Formal Parameters

Formal parameters play a fundamental role in OPL: they are used in aggregate operators, queries, and `forall` statements. This chapter reviews them in more detail. Section 6.1 describes the basic formal parameters, Section 6.2 shows how to use tuples of formal parameters, Section 6.3 studies how tuples of parameters can be used for filtering, and Section 6.4 discusses a number of modeling issues that arise when using parameters. Figure 6.1 depicts the syntax of parameters in OPL.

6.1 Basic Formal Parameters

The simplest formal parameter has the form

```
p in S
```

where `p` is the formal parameter and `S` is the set from which `p` takes its values. The set `S` can be an integer range, as in

```
int s = sum(i in 1..n) i*i;
```

an enumerated type, as in

```
enum Products ...;
float+ cost[Products] = ...;
float+ maxCost = max(p in Products) cost[p];
```

or a finite set, as in

```
enum Cities ...;
struct Connection {
    Cities orig;
    Cities dest;
};
{Connection} connections = ...;
float+ cost[connections] = ...;
float+ maxCost = max(r in connections) cost[r];
```

It is sometimes desirable to restrict the range of the formal parameters using conditions. The formal parameter then takes the form

```
p in S : condition
```

and assigns to `p` all elements of `S` satisfying the condition. For instance, the excerpt

---

$\langle ListParameter \rangle$ → $\langle Parameter \rangle$
→ $\langle Parameter \rangle$, $\langle ListParameter \rangle$
→ $\langle Parameter \rangle$ & $\langle ListParameter \rangle$

$\langle Parameter \rangle$ → $\langle Parameter \rangle$
→ $\langle Object \rangle^+$ in $\langle Bounds \rangle$
→ $\langle Ordered \rangle \langle Object \rangle^+$ in $\langle Bounds \rangle$
→ $\langle Object \rangle^+$ in $\langle Bounds \rangle$: $\langle Relation \rangle$
→ $\langle Ordered \rangle \langle Object \rangle^+$ in $\langle Bounds \rangle$:$\langle Relation \rangle$

$\langle Object \rangle$ → $\langle Id \rangle$ | < $\langle Id \rangle^+$ >
$\langle Bounds \rangle$ → $\langle Composite \rangle$ | $\langle Range \rangle$

Figure 6.1: The Parameters Syntax.

```
forall(i in 1..8)
    forall(j in 1..8 : i < j)
        queen[i] <> queen[j];
```

enforces the constraint queen[i] <> queen[j] for all $1 \leq i < j \leq 8$. Several parameters can often be combined together to produce more compact statements. For instance, the declaration

```
int s = sum(i,j in 1..n) i*j;
```

is equivalent to

```
int s = sum(i in 1..n) sum(j in 1..n) i*j;
```

which is less readable. The declaration

```
int s = sum(i in 1..n & j in 1..m) i*j;
```

where the ranges of i and j are different, is equivalent to

```
int s = sum(i in 1..n, j in 1..m) i*j;
```

where a comma has been substituted for &, and to

```
int s = sum(i in 1..n) sum(j in 1..m) i*j;
```

These parameters can, of course, be subject to conditions. The excerpt

```
forall(i,j in 1..8 :  i < j)
   queen[i] <> queen[j];
```

is equivalent to

```
forall(i in 1..8 & j in 1..8 :  i<j)
   queen[i] <> queen[j];
```

and to the excerpt shown earlier in this chapter. The even more compact form

```
forall(ordered i,j in 1..8)
   queen[i] <> queen[j];
```

is equivalent to the excerpts just shown and illustrates a functionality, i.e., ordered, often useful in practical applications. Indeed, in many applications one is interested, given a set S, to state constraints or conditions over all pairs (i,j) of elements of S satisfying $i < j$ in the ordering associated with S. Note that S may be a range, an enumerated type, or any finite set, since these are all ordered. For instance, the excerpt

```
enum Tasks ...;
enum Resources ...;
{Tasks} res[Resources] = ...;
...
int duration[Tasks] = ...;
var int start[Tasks] in 0..10000;
solve {
   forall(r in Resources & ordered t1,t2 in res[r])
      start[t1] >= start[t2] + duration[t2] \/
      start[t2] >= start[t1] + duration[t1];
   ...
};
```

illustrates this functionality on an enumerated type Task. The quantifier considers all pairs (t1,t2) in the set of tasks res[r] such that t1 < t2.

6.2 Tuples of Parameters

OPL also allows tuples of formal parameters to appear in aggregate operators, forall statements, and queries. Consider the excerpt

```
enum Tasks ...;
struct Precedence {
    Tasks before:
    Tasks after;
};
{Precedence} Prec = ...;
int duration[Tasks] = ...;
var int start[Tasks] in 0..maxTime;
solve {
    forall(p in Prec)
        start[p.after] >= start[p.before] + duration[p.before];
};
```

which states precedence constraints between tasks. The constraint declaration requires explicit accesses to the fields of the record to state the constraints. In addition, the field `before` is accessed twice. A more elegant way to state the same constraint is to use a tuple of formal parameters, as in

```
forall(<b,a> in Prec)
    start[a] >= start[b] + duration[b]
```

precluding the need to access the record fields explicitly. The tuple `<b,a>` in the `forall` quantifier contains two parameters that are given the values of the fields of each record in `Prec` successively. More generally, an expression

```
p in S
```

where `S` is a set of records containing n fields, can be replaced by a formal parameter expression

```
<p1,...,pn> in S
```

that contains n formal parameters. Each time a record `r` is selected from `S`, its fields are assigned to the corresponding formal parameters. This functionality is often useful in producing more readable models.

6.3 Filtering in Tuples of Parameters

OPL also enables simple equality constraints to be factorized inside the tuples, which is important in obtaining more readable and efficient models. Consider, for instance, a transportation problem where products must be shipped from a set of cities to another set of cities. The model may include a constraint specifying that the total shipments for all products transported along a connection may not exceed a specified limit. This can be expressed by a constraint

6. Formal Parameters

```
forall(c in connections)
   sum(<p,co> in routes :  c = co) trans[<p,c>] <= limit;
```

This constraint states that the total products shipped along each connection c is not greater than limit. This statement is particularly inefficient: OPL must scan the entire set routes to select the tuples involving each connection. The constraint would be stated more efficiently as follows

```
forall(c in connections)
   sum(<p,c> in routes) trans[<p,c>] <= limit;
```

In this last constraint, the tuple <p,c> contains one new parameter p and uses the previously defined parameter c. Since the value of c is known, OPL uses it to index the set routes, avoiding a complete scanning of the set routes.

6.4 Modeling Issues

It is useful at this point to discuss some modeling issues involving sparsity, tuples of parameters, and filtering. An application can often be described by various models that may exhibit fundamentally different performance in terms of memory and computing time. This is particularly important for large-scale linear models.

Consider again the transportation problem in which the shipments of products between each pair of cities may not exceed a given limit. Statement 6.1 shows a simple model for this problem, which implicitly assumes that all cities are connected and that all products may be shipped between two cities. It is thus not appropriate for large-scale problems where only a fraction of the cities are connected. A small dataset can easily illustrate the issue: Consider the set of cities

{Amsterdam,Antwerpen,Bergen,Bonn,Brussels,Cassis,London,Madrid,Milan,Paris}

and the set of products {Godiva,Leonidas,Neuhaus}. There are three hundred ways of shipping a product from a city to another. However, only a small fraction of these may be explored in the application and Figure 6.2 displays a possible subset. Using Statement 6.1 would induce a substantial loss in (memory and time) efficiency. The rest of this section explores how to exploit this sparsity.

6.4.1 A First Attempt

A first attempt at exploiting the sparsity available in a large-scale transportation problem consists of representing the data as a set routes of records of type

```
struct Route { Prod p, Cities o, Cities d, },
```

```
enum Cities ...;
enum Products ...;
float+ limit = ...;
float+ supply[Products,Cities] = ...;
float+ demand[Products,Cities] = ...;
assert forall(p in Products)
   sum(o in Cities) supply[p,o] = sum(d in Cities) demand[p,d];
float+ cost[Products,Cities,Cities] = ...;

var float+ trans[Products,Cities,Cities];
minimize
   sum(p in Products & o,d in Cities) cost[p,o,d] * trans[p,o,d]
subject to {
   forall(p in Products & o in Cities)
      sum(d in Cities) trans[p,o,d] = supply[p,o];
   forall(p in Products & d in Cities)
      sum(o in Cities) trans[p,o,d] = demand[p,d];
   forall(o, d in Cities)
      sum(p in Products) trans[p,o,d] <= limit;
};
```

Statement 6.1: A Simple Transportation Model (transp1.mod).

<Godiva,Brussels,Paris>	Godiva,Brussels,Bonn>	<Godiva,Amsterdam,London>
<Godiva,Amsterdam,Milan>	<Godiva,Antwerpen,Madrid>	<Godiva,Antwerpen,Bergen>
<Neuhaus,Brussels,Milan>	<Neuhaus,Brussels,Bergen>	<Neuhaus,Amsterdam,Madrid>
<Neuhaus,Amsterdam,Cassis>	<Neuhaus,Antwerpen,Paris>	<Neuhaus,Antwerpen,Bonn>
<Leonidas,Brussels,Bonn>	<Leonidas,Brussels,Milan>	<Leonidas,Amsterdam,Paris>
<Leonidas,Amsterdam,Cassis>	<Leonidas,Antwerpen,London>	<Leonidas,Antwerpen,Bergen>

Figure 6.2: A Sparse Dataset for a Transportation Problem

6. Formal Parameters

```
enum Cities ...;
enum Products ...;
float+ limit = ...;

struct Route { Products p; Cities o; Cities d; };
{Route} routes = ...;
struct Supply { Products p; Cities o; };
{Supply} Supplies = { <p,o> | <p,o,d> in routes };
float+ supply[Supplies] = ...;
struct Demand { Products p; Cities d; };
{Demand} Demands = { <p,d> | <p,o,d> in routes };
float+ demand[Demands] = ...;
float+ cost[routes] = ...;

{Cities} orig[p in Products] = { o | <p,o,d> in routes };
{Cities} dest[p in Products] = { d | <p,o,d> in routes };
assert forall(p in Products)
   sum(o in orig[p]) supply[<p,o>] = sum(d in dest[p]) demand[<p,d>];
var float+ trans[routes];

minimize
   sum(l in routes) cost[l] * trans[l]
subject to {
   forall(p in Products & o in orig[p])
      sum(d in dest[p]) trans[<p,o,d>] = supply[<p,o>];
   forall(p in Products & d in dest[p])
      sum(o in orig[p]) trans[<p,o,d>] = demand[<p,d>];
   forall(o,d in Cities)
      sum(<p,o,d> in routes) trans[<p,o,d>] <= limit;
};
```

Statement 6.2: A Sparse Transportation Model: First Attempt (transp2.mod).

The array `cost` and `trans` can then be indexed with this set. A model based on this idea appears in Statement 6.2. The data for the supplies and demands are also represented in a sparse way by projecting the set `routes` to obtain their index sets. In addition to that, the model also pre-computes, in a generic way, the cities `orig[p]` that can ship product `p` and the cities `dest[p]` that can receive product `p`. Most of the resulting model is elegant and efficient. Unfortunately, the constraint

```
forall(o in Orig & d in Dest)
   sum(<p,o,d> in routes) trans[<p,o,d>] <= limit;
```

is not particularly efficient because it does not exploit the structure of the application. Indeed, the `forall` statement iterates not over actual connections but rather over all pairs of cities. In addition, the aggregate operator

```
sum(<p,o,d> in routes) trans[<p,o,d>] <= limit;
```

cannot exploit the "connection" structure to obtain all products of a connection, since `o` and `d` are separate entities.

6.4.2 A Better Model

The application can be modeled more effectively by closely reflecting the structure of the application. Figure 6.3 gives a statement illustrating this principle. The main novelty is the explicit representation of connections and the fact that a route is now simply the association of a connection and a product. Connections are also computed automatically from routes. The rest of the model is generally similar but reflects the new data organization. The most interesting change is the capacity constraint, which becomes

```
forall(c in connections)
   sum(<c,p> in routes) trans[<c,p>] <= limit;
```

This constraint is much more efficient than in the previous model. First, it iterates over the routes, not over all pairs of cities. Second, the aggregate operator `sum` uses parameter `c` to index the set `routes`, retrieving the relevant products effectively.

```
enum Cities ...;
enum Products ...;

struct Connection { Cities o; Cities d; };
struct Route { Connection e; Products p; };
struct Supplier { Products p; Cities o; };
struct Customer { Products p; Cities d; };

{Route} Routes = ...;
{Connection} Connections = { c | <c,p> in Routes };
{Supplier} Suppliers = { <p,c.o> | <c,p> in Routes };
float+ supply[Suppliers] = ...;
{Customer} Customers = { <p,c.d> | <c,p> in Routes };
float+ demand[Customers] = ...;
float+ lim = ...;
float+ cost[Routes] = ...;
{Cities} orig[p in Products] = { c.o | <c,p> in Routes };
{Cities} dest[p in Products] = { c.d | <c,p> in Routes };

assert forall(p in Products)
   sum(o in orig[p]) supply[<p,o>] = sum(d in dest[p]) demand[<p,d>];

var float+ trans[Routes];

minimize
   sum(r in Routes) cost[r] * trans[r]
subject to {
   forall(p in Products & o in orig[p])
      sum(d in dest[p]) trans[<<o,d>,p>] = supply[<p,o>];
   forall(p in Products & d in dest[p])
      sum(o in orig[p]) trans[<<o,d>,p>] = demand[<p,d>];
   forall(c in Connections)
      sum(<c,p> in Routes) trans[<c,p>] <= lim;
};
```

Statement 6.3: A Sparse Transportation Model: Second Attempt (transp3.mod).

7 Search

One of the main novelties of OPL is its ability to specify search procedures. This support is fundamental for hard combinatorial optimization problems, where search is necessary. Indeed, constraint-solving algorithms are typically incomplete and cannot solve optimization problems alone, except in some specific classes of applications such as linear programming. The constraint-solving algorithms usually solve relaxations of the problem and are complemented by search procedures. Traditionally, modeling languages and most mathematical programming packages hide these search procedures from users, who can control them only through a set of parameters. OPL also has a number of default search procedures. In addition, it offers the ability to specify search procedures tailored to the application at hand. This functionality, at the core of traditional constraint programming languages, can lead to significant improvements in performance by incorporating special-purpose heuristics in the model. It is important to emphasize that the search procedure is optional and may be partial (e.g., not all variables are given a value). OPL simply applies its default search procedures once the user-defined search has been executed. The syntax of search procedures is given in Figure 7.1 and this chapter reviews the various constructs available.

7.1 The Try Instruction

Many applications use search algorithms that generate values for the variables. OPL has a number of constructs to express these procedures. Its most basic control structure is the **try** instruction, for example

```
try
      x = 1
    |
      x = 2
endtry;
```

The **try** instruction is nondeterministic and specifies a number of alternatives to be explored. The instruction in the example tells OPL to assign x to either 1 or 2; it succeeds if one of these alternatives is consistent with the constraint store and it fails otherwise. Operationally, OPL tries the first alternative and adds the constraint x = 1 to the constraint store. When the constraint is added to the store, OPL either detects a failure (i.e., the constraint is not consistent with the store), in which case the next alternative is selected, or succeeds, in which case OPL moves on to the next instructions, i.e., the instructions following the try if any. The execution of these subsequent instructions may fail, in which case the other alternative to the **try** (i.e. x = 2) is considered. For instance, the model

⟨Search⟩	→	search ⟨Choice⟩ ;
⟨Choice⟩	→	
	→	try ⟨TryChoice⟩ endtry
	→	tryall (⟨Formal⟩ [: ⟨Relation⟩] [⟨COrder⟩]) ⟨Choice⟩
	→	forall (⟨Formal⟩ [: ⟨Relation⟩] [⟨COrder⟩]) ⟨Choice⟩
	→	if ⟨Relation⟩ then ⟨Choice⟩ [else ⟨Choice⟩] endif
	→	while ⟨Relation⟩ do ⟨Choice⟩
	→	select (⟨Formal⟩ [: ⟨Relation⟩] [⟨COrder⟩]) ⟨Choice⟩
	→	let ⟨Id⟩ = ⟨Expr⟩ in ⟨Choice⟩
	→	once ⟨Choice⟩
	→	when ⟨Relation⟩ do ⟨Choice⟩
	→	onValue ⟨Expr⟩ do ⟨Choice⟩
	→	onRange ⟨Expr⟩ do ⟨Choice⟩
	→	onDomain ⟨Expr⟩ do ⟨Choice⟩
	→	⟨Relation⟩
	→	{ ⟨Choice⟩$^+$ }
	→	⟨Generate⟩ (⟨Composite⟩)
	→	⟨Branch⟩ (⟨Composite⟩)
	→	rank | rankLocal | rankGlobal | rank (⟨Composite⟩)
	→	setTimes | setTimes (⟨Composite⟩)
	→	assignAlternatives
	→	tryRankFirst (⟨Composite⟩ , ⟨Composite⟩)
	→	tryRankLast (⟨Composite⟩ , ⟨Composite⟩)
⟨TryChoice⟩	→	⟨Choice⟩
	→	⟨Choice⟩ | ⟨TryChoice⟩
⟨Formal⟩	→	⟨Object⟩ in ⟨Bounds⟩
⟨COrder⟩	→	ordered by increasing ⟨Exp⟩
	→	ordered by decreasing ⟨Exp⟩
	→	ordered by increasing < ⟨Expr⟩$^+$ >
	→	ordered by decreasing < ⟨Expr⟩$^+$ >
⟨Generate⟩	→	generate | generateSize | generateSeq | generateMin | generateMax
⟨Branch⟩	→	branch | branchLow | branchUp | split | splitLow | splitUp

Figure 7.1: The Syntax of Search Procedures.

7. Search

```
var int x in 1..5;
solve {
    x <> 1;
};
search {
    try x = 1 | x = 2 | x = 3 | x = 4 | x = 5 endtry;
};
```

adds the constraint x = 1, which produces a failure. It then adds the constraint x = 2, which succeeds and produces the solution x = 2. If another solution is requested, OPL backtracks, selects the next alternative x = 3, and produces the solution x = 3. The other solutions are produced in a similar fashion.

It is interesting to study what happens when several **try** instructions are used in sequence. Consider the model

```
var int x in 0..1;
var int y in 0..1;
solve {
    x + y <=1;
};
search {
    try x = 1 | x = 0 endtry;
    try y = 1 | y = 0 endtry;
};
```

Here OPL considers the first **try** instruction and adds the constraint x = 1. It then moves to the next instruction and adds the constraint y = 1, which results in a failure. OPL then backtracks to the second alternative and produces the solution

```
Solution [1]
  x = 1
  y = 0
```

If additional solutions are requested, OPL backtracks to the second **try** instruction. Since no alternatives remain, it then backtracks to the first **try** instruction and restores the constraint store to the exact same state as when the **try** instruction was first executed. The constraint store at that point only contains the constraint x + y <= 1. OPL then selects the second alternative x = 0, moves to the next instruction, selects the first alternative, and produces the solution

```
Solution [2]
```

```
x = 0
y = 1
```

7.2 The Tryall Instruction

The `try` instruction is convenient when the set of alternatives is small and does not depend on the input data. The `tryall` instruction can be viewed as a compact and dynamic representation of the `try` instruction. For instance, the instruction

```
tryall(i in 1..5)
    x = i;
```

is equivalent to the instruction

```
try x = 1 | x = 2 | x = 3 | x = 4 | x = 5 endtry;
```

It first assigns 1 to parameter `i` and adds the constraint `x = i` to the constraint store. If it succeeds, the execution proceeds to the next instruction. Otherwise, the instruction assigns the value 2 to `i` and adds the constraint `x = i` again. This time, however, the constraint corresponds to `x = 2`. The `tryall` instruction can also specify the order in which the alternatives should be tried. For instance, the instruction

```
tryall(i in 1..5 ordered by increasing i)
    x = i;
```

tries the values for `i` in the order 1, 2, ..., 5, while the instruction

```
tryall(i in 1..5 ordered by decreasing i)
    x = i;
```

tries the values in the order 5, 4, ..., 1. Note that the ordering can be specified by an arbitrary expression that may contain some reflective functions. Finally, it is useful to mention that the `tryall` instruction may include a condition, as in

```
tryall(i in 1..10 : i mod 3 = 0)
    x = i;
```

which is equivalent to

```
try x = 3 | x = 6 | x = 9 endtry;
```

The condition can be used to filter out some undesired elements.

7.3 Quantifiers

The other fundamental choice instruction is a universal quantifier (or iterative statement) that is similar in spirit to the universal quantifier of constraint specifications. However, the formal parameters allowed in the `forall` statement of search procedures (as well as in other constructs such as `tryall` and `select`) are a subset of those available for constraints. This restriction is motivated by the need to define dynamic orderings, which are often important in practice.

To illustrate the `forall` instruction, consider the following statement, which solves the four-queens problem and specifies a search procedure:

```
var int queen[1..4] in 1..4;
solve {
   forall(ordered i, j in 1..4) {
      queen[i] <> queen[j];
      queen[i] + i <> queen[j] + j;
      queen[i] - i <> queen[j] - j;
   }
};
search {
   forall(i in 1..4)
      tryall(v in 1..4)
         queen[i] = v;
};
```

The search procedure specifies considering each queen in turn (i.e., `queen[1]`, ..., `queen[4]`) and assigning them the values `1..4` nondeterministically. This search procedure is in fact equivalent to

```
search {
    try queen[1] = 1 | queen[1] = 2 | queen[1] = 3 | queen[1] = 4 endtry;
    try queen[2] = 1 | queen[2] = 2 | queen[2] = 3 | queen[2] = 4 endtry;
    try queen[3] = 1 | queen[3] = 2 | queen[3] = 3 | queen[3] = 4 endtry;
    try queen[4] = 1 | queen[4] = 2 | queen[4] = 3 | queen[4] = 4 endtry;
};
```

When the model is executed, the search procedure assigns the value 1 to `queen[1]`. It then tries assigning a value to `queen[2]` but all values fail. The search procedure then backtracks and assigns the value 2 to `queen[1]`. It then tries the values `1..4` for `queen[2]`. Values `1..3` lead to failures but value 4 succeeds and assigns values 1 and 3 to `queen[3]` and `queen[4]`, respectively. The last two `tryall` instructions succeed for their first and third alternatives.

The `forall` instructions can also be ordered dynamically, which is an important functionality for many applications. For instance, the search procedure

```
search {
    forall(i in 1..8 ordered by increasing dsize(queen[i]))
        tryall(v in 1..8)
            queen[i] = v;
};
```

selects first the queen with the fewest values in its domain. It assigns a value to this queen and, in case of success, selects the queen which has the smallest domain among the remaining queens. Note that the domains used in selecting the next queen to assign are computed with respect to the constraint store after the first assignment. In other words, the ordering is dynamic.

The `forall` instructions can also be ordered by tuples of expressions. For instance, the search procedure

```
search {
    forall(i in 1..8 ordered by increasing <dsize(queen[i]),dmin(queen[i])>)
        tryall(v in 1..8)
            queen[i] = v;
};
```

selects the queen that has the smallest domain and, in case of ties, the queen that has the smallest value in its domain. In other words, tuples of expressions are compared using the lexicographic ordering

$\langle a, b \rangle \leq \langle c, d \rangle$ if $a < c$ or $a = c$ & $b \leq d$.

The `forall` instructions can also be subject to conditions. The search procedure

```
search {
    forall(i in 1..10 :  i mod 2 = 0)
        tryall(v in 1..10)
            x[i] = v;
};
```

generates values for the even indices of x. It is useful to emphasize at this point that OPL supports both variable ordering (in `forall` instructions) and value ordering (in `tryall` instructions), two techniques commonly used in constraint satisfaction problems.

7.4 Sequencing Choices

As should be clear, search procedures in OPL are sequences of instructions. The search procedure

```
search {
    try x = 1 | x = 2 endtry;
    try y = 1 | y = 2 endtry;
};
```

executes the first **try** instruction and, in case of success, proceeds to the second **try** instruction. Of course, the sequences may contain arbitrary choice instructions, including **forall** and **tryall** instructions. For instance, the search procedure

```
search {
    forall(i in 1..3)
        tryall(v in 1..3)
            x[i] = v;
    forall(j in 1..3)
        tryall(w in 1..3)
            y[j] = w;
};
```

specifies a sequence of two **forall** instructions. This search procedure could be written, by changing only the names of the formal parameters, as

```
search {
    forall(i in 1..3)
        tryall(v in 1..3)
            x[i] = v;
    forall(i in 1..3)
        tryall(v in 1..3)
            y[i] = v;
};
```

since the scope of the parameters in **forall** and **tryall** statements is the body of their instructions. Note that the above search procedure is fundamentally different from

```
search {
    forall(i in 1..3) {
        tryall(v in 1..3)
            x[i] = v;
```

```
            tryall(v in 1..3)
                y[i] = v;
        };
    };
```

Indeed, the former search procedure is equivalent to

```
search {
    try x[1] = 1 | x[1] = 2 | x[1] = 3 endtry;
    try x[2] = 1 | x[2] = 2 | x[2] = 3 endtry;
    try x[3] = 1 | x[3] = 2 | x[3] = 3 endtry;
    try y[1] = 1 | y[1] = 2 | y[1] = 3 endtry;
    try y[2] = 1 | y[2] = 2 | y[2] = 3 endtry;
    try y[3] = 1 | y[3] = 2 | y[3] = 3 endtry;
};
```

while the latter is equivalent to

```
search {
    try x[1] = 1 | x[1] = 2 | x[1] = 3 endtry;
    try y[1] = 1 | y[1] = 2 | y[1] = 3 endtry;
    try x[2] = 1 | x[2] = 2 | x[2] = 3 endtry;
    try y[2] = 1 | y[2] = 2 | y[2] = 3 endtry;
    try x[3] = 1 | x[3] = 2 | x[3] = 3 endtry;
    try y[3] = 1 | y[3] = 2 | y[3] = 3 endtry;
};
```

7.5 Conditional Choices

Conditionals can also be used in OPL to make different choices according to the truth value of a condition. For instance, the search procedure

```
search {
    forall(i in 1..10)
        if i mod 2 = 0 then
            tryall(v in 1..10 ordered by increasing i)
                x[i] = v
        else
            tryall(v in 1..10 ordered by decreasing i)
```

7. Search

```
            x[i] = v
     endif;
}
```

uses a different value ordering for the variables with even and odd indices. Note that the condition in the conditional statement must not contain variables except, of course, in reflective functions.

7.6 The While Instruction

OPL supports a `while` instruction that executes a choice statement while a condition is satisfied. This construct is useful in a variety of circumstances, e.g., to generate constraints until a variable is bound or to order activities in a unary resource until the resource is fully ranked. For instance, the search procedure

```
search {
    while not bound(x) do
        try x = dmin(x) | x <> dmin(x) endtry;
};
```

nondeterministically chooses between assigning to x the minimum value in its domain or removing this minimum value from its domain. Such choices are iterated until x is bound. Note that this search procedure may have an undesired side-effect in optimization problems, since the minimum value of x may be removed by the branch and bound before execution of the instruction x <> dmin(x). An appropriate alternative is shown in Section 7.8.

7.7 The Select Instruction

Another useful construct supported by OPL is the `select` instruction, which selects data satisfying some conditions. This is especially useful in conjunction with the `while` instruction. For instance, the following search procedure, which assumes that x is a one-dimensional array of variables,

```
search {
    while not bound(x) do
        select(i in 1..10 : not bound(x[i]))
            tryall(v in 1..10) x[i] = v;
};
```

selects an index i such that x[i] is not bound, assigns a value to x[i], and iterates the process until all variables in x are bound or no such assignment exists. Note that the `select` instruction only selects one value and commits to it. The `select` instruction can also include an ordering, in which case the elements are selected according to the specified ordering. For instance, the search procedure

```
search {
    while not bound(x) do
        select(i in 1..10 such that not bound(x[i]))
                        ordered by increasing dsize(x[i]))
            tryall(v in 1..10) x[i] = v;
};
```

chooses, at each iteration, the index of the non-bound variable that has the smallest domain.

7.8 The Let Statement

Common expressions can be factored using the `let` construct. Consider the search procedure

```
search {
    while not bound(x) do
        try x = dmin(x) | x <> dmin(x) endtry;
};
```

presented earlier in this chapter. It could be rewritten to factor the expression `dmin(x)` as

```
search {
    while not bound(x) do
        let m = dmin(x) in
            try x = m | x <> m endtry;
};
```

The `let` statement declares the parameter `m` and initializes it with the value of the expression `dmin(x)`. Parameter `m` can then be used in the body of the `let` statement in place of the expression. This last formulation is more appropriate, since it guarantees that the value `m` that is removed by the instruction `x <> m` is the same as the value that is tried in `x = m`. This is not necessarily the case in the previous formulation.

7.9 The Once Statement

The instruction `once` is useful when only one solution to a goal is desired. A natural application of this construct is described in Chapter 11. The instruction `once C` searches for a solution to `C` and, in case of success, commits to the solution. In other words, no backtracking occurs in `C` as soon as its first solution has been found. To understand this behavior of once, it is useful to contrast the statement

```
var int x in 0..10;
solve x <> 1;
search {
   once {
      tryall(i in 0..10)
         x = i;
      x <> 0;
   };
};
```

which returns the unique solution

```
Solution [1]
  x = 2
```

and the statement

```
var int x in 0..10;
solve x <> 1;
search {
   once {
      tryall(i in 0..10)
         x = i;
   };
   x <> 0;
};
```

which fails. This second statement fails because the tryall instruction commits after its first solution (which adds the constraint x = 0). The constraint x <> 0 then fails and there is no choice point left to explore. Note that instruction once is procedural in nature and should be used with care (e.g., when it is known that the solution exists and is appropriate). Again, a natural application of this instruction is presented in Chapter 11.

7.10 Constraints

Constraints are of course the building blocks of search procedures. The examples considered earlier in this section use only equations, but arbitrary constraints can be used in the search procedures. For instance, the search procedure

```
search {
    forall(<f,s> in Disjunctions)
       try
          start[f] >= end[s]
       |
          start[s] >= end[f]
       endtry;
};
```

illustrates the use of inequalities in the various alternatives of a **try** statement.

7.11 Data-Driven Constructs

OPL also supports a variety of data- or constraint-driven constructs in the spirit of concurrent constraint programming. The key idea here is to execute an instruction whenever some condition becomes true or some event occurs. The resulting computational model can thus be viewed as a set of processes communicating through the constraint store. Consider, for instance, the search procedure

```
search {
    forall(w in Warehouses)
       when open[w] = 0 do
          forall(c in Customers)
             supply[c] <> w;
    forall(w in Warehouses)
       try open[w] = 0 | open[w] = 1 endtry;
};
```

The instruction

```
when open[w] = 0 do
    forall(c in Customers)
       supply[c] <> w;
```

is a data-driven construct. When first encountered, it tests if the constraint open[w] = 0 holds in the constraint store. So, the body

7. Search

```
forall(c in Customers)
    supply[c] <> w;
```

is executed. If its negation `open[w] <> 0` holds, the instruction succeeds and does nothing. Otherwise (e.g., if `open[w]` is not yet known in the constraint store), the instruction suspends. As soon as more information on `open[w]` is available in the constraint store, the instruction is reconsidered to determine whether its body can now be executed. This happens, for instance, when `open[w]` is assigned a value later on in the above choice procedure. It is useful to think of these data-driven constructs as processes that wait until some condition or event take places before being executed. In the above procedure, there are as many processes as there are warehouses. Of course, the OPL implementation has efficient ways of processing these constructs.

OPL offers several other data-driven constructs in addition to **when**. The construct **onValue** waits until an expression has a fixed value to execute its body. For instance, the search procedure

```
search {
    forall(w in Warehouses)
        when open[w] = 0 do
            forall(c in Customers)
                supply[c] <> w;
};
```

is equivalent to

```
search {
    forall(w in Warehouses)
        onValue open[w] do
            if open[w] = 0 then
                forall(c in Customers)
                    supply[c] <> w
            endif;
};
```

This **onValue** instruction suspends until `open[w]` has been assigned a value. As soon this happens, the body

```
if open[w] = 0 then
    forall(c in Customers)
        supply[c] <> w
endif;
```

is executed. The `when` and `onValue` constructs execute their body at most once. In contrast, the `onRange` and `onDomain` constructs execute their bodies many times in general. An `onRange` instruction, say

```
onRange queen[1] do
   queen[2] <> dmin(queen[1]);
```

executes the instruction

```
queen[2] <> dmin(queen[1])
```

each time the range of `queen[1]` is modified (i.e., each time its minimum or maximum value is removed). This instruction is very procedural and its use is discouraged in general. The `onDomain` instruction is essentially similar but executes its body whenever the domain of the expression is modified.

7.12 Predefined Search Strategies

In addition to the above general constructs, OPL provides a number of predefined functions to define search procedures. This section considers in turn generation, branching, and scheduling search procedures.

7.12.1 Generation Procedures

Generation procedures receive a discrete variable, or an arbitrary array of discrete variables, and generate values for all these variables. For instance, the four-queens problem described earlier can be written as

```
var int queen[1..4];
solve {
   forall(ordered i, j in 1..4) {
      queen[i] <> queen[j];
      queen[i] + i <> queen[j] + j;
      queen[i] - i <> queen[j] - j;
   };
};
search {
   forall(i in 1..4)
      generate(queen[i]);
};
```

The search procedure in this statement can even be replaced by the procedure

```
search {
    generate(queen);
};
```

which also generates values for all queens. OPL has various generation instructions. The instruction `generateSize` generates values for the variables in its arguments by generating values for the variable with the smallest domain first, while the instructions `generationMin` and `generationMax` consider first the variable with the smallest and the largest value in its domain, respectively. In the current implementation, `generate` is equivalent to `generateSize`. Instruction `generateSeq` considers the variables in the fixed order of the array.

7.12.2 Branching Instructions

In integer programming problems, it is often appropriate to design a search procedure splits the domain of a variable in two parts and explores both parts nondeterministically. The splitting point is generally the value of the variable in the linear relaxation. Such a procedure could be implemented as follows:

```
while not bound(x) do
    let v = simplexValue(x) in
    let l = floor(v) in
    let v = ceil(v) in
        try
            x <= l
        |
            x >= v
        endtry;
    endif
```

This search procedure is supported directly by the instruction `splitLow`. The instruction `splitUp` is similar, except that the order of the alternatives in the `try` instructions is permuted; the instruction `split` is similar but the order is chosen by OPL. The instructions `branch`, `branchLow`, and `branchUp` generalize the `split` instructions to arrays of variables. For instance, assuming that x is a one-dimensional array of integer variables, the instruction `branchLow(x)` could be written as

```
while not bound(x) do
    select(i in Index : not bound(x[i]))
        splitLow(x[i]);
```

Of course, these instructions can be applied to arrays of arbitrary dimensions.

7.13 Choices in Scheduling

OPL also supports some predefined search strategies for scheduling applications. Scheduling applications are discussed in Chapter 11 and it may be helpful to read that chapter before studying this section.

Rank Instructions The `rank` instructions are used for unary resources. Their goal is to order all activities on the unary resources. For instance, the instruction

`rank(r)`

ranks all activities on the unary resources `r`. After its execution, all the activities of the resource are totally ordered by precedence constraints and cannot overlap in time. In addition, OPL provides three ranking instructions that apply to all unary resources: `rank`, `rankLocal`, and `rankGlobal`. They differ in the resources chosen to be ranked next. The basic idea in these instructions is to select the resource that is most constrained. The procedure `rankGlobal` uses the *global slack* of the resource that considers all activities that are not yet ranked, computes the earliest starting date `s`, the latest finishing date `e`, and the total duration `d` of all these activities, and returns `(e - s) - d`. This produces an approximation of the tightness of the resource at a given computation point. The procedure `rankLocal` uses the *local slack* of the unary resource, another measure of tightness that is more precise but more expensive to compute. It entails computing the global slack for a variety of subsets of the unranked activities. In addition to these coarse-grained instructions, OPL provides two instructions `tryRankFirst` and `tryRankLast` that receive, as arguments, an activity and a resource. Instruction `tryRankFirst` tries to rank the activity before all non-ranked activities on the resource. On backtracking, it specifies that the activity cannot be ranked first. Instruction `tryRankLast` is similar but tries to rank the activity last on resources.

SetTimes Instructions The instruction `setTimes` is useful in some classes of scheduling problems with discrete resources. Given an array `a` of activities, `setTimes(a)` assigns starting dates to all activities in `a`. This instruction should not be applied blindly: there are problems for which it can miss solutions (e.g., when there are negative distance constraints of the form `a.start >= b.end - 3`). However, for problems with discrete resourses, positive distance constraints, and activities with fixed duration, it may considerably improve efficiency over a more naive strategy.[1] Under this hypothesis, the basic idea behind `setTimes` can be explained as follows. OPL selects the earliest starting date `d` of all activities and chooses an activity that can be scheduled at date `d`. It then

---

[1] Of course, it is often possible to make choices so as get to a position where `setTimes` can be applied naturally.

creates a choice point with two alternatives: the first alternative schedules the activity at date d, while the second alternative *postpones* the activity. The process is then repeated for all activities that are not yet scheduled or not postponed. A postponed activity is reconsidered whenever its starting date is updated. Note that OPL also supports an instruction `setTimes` with no argument that applies to all activities in the model.

Instruction assignAlternatives When alternative resources are used in a model, each activity using the alternative resources must be assigned a resource from its set of unary resources. OPL supports this by providing a nondeterministic instruction `assignAlternatives` that, when it succeeds, guarantees that a resource has been selected for each constraint using an alternative resource. The rest of the search procedure can then specify how to solve the resulting problem. Note also that constraints such as `activityHasSelectedResource(a,s,u)` can be used to defined a strategy tailored to the problem at hand.

8 Display

An OPL model may also specify how to display the results. Although this information is partly subsumed by the browsing facility of OPL STUDIO, the display information, and the ability to specify postprocessing of data, adds flexibility and convenience. This chapter reviews this functionality in detail; the syntax of the display instructions is given in Figure 8.1.

8.1 Displaying Data

By default, OPL STUDIO displays the values of all variables and activities in an optimal solution (optimization problems) or in a solution (decision problems). This default can be overwritten, in which case only the value of the objective function (in optimization problems) and the fact that there exists a solution is displayed. However, in practice, it is often useful to provide a finer control on the display. The simplest display instructions specify the data to be displayed by listing the appropriate identifiers. Figure 8.1 shows a transportation model where the instruction

```
display trans;
```

specifies that the array `trans` must be displayed. The display produced by the model has the form

```
Optimal Solution at Cost:   206400.0000

  trans[Bruxelles,Paris] = 0.0000
  trans[Bruxelles,Lille] = 0.0000
  trans[Bruxelles,Nancy] = 0.0000
  trans[Bruxelles,Kocln] - 0.0000
  trans[Bruxelles,Amsterdam] = 600.0000
  trans[Bruxelles,London] = 1100.0000
  trans[Bruxelles,Calais] = 0.0000
  trans[Louvain,Paris] = 0.0000
  trans[Louvain,Lille] = 1200.0000
  trans[Louvain,Nancy] = 600.0000
  trans[Louvain,Koeln] = 400.0000
  trans[Louvain,Amsterdam] = 0.0000
  trans[Louvain,London] = 0.0000
  trans[Louvain,Calais] = 600.0000
  ...
```

Display instructions may specify any data declared in the model, although variables and activities are obviously most interesting.

⟨*Display*⟩   →   display ⟨*DisExpr*⟩ ;
            →   display (⟨*ListParameter*⟩) ⟨*DisExpr*⟩ ;
            →   ⟨*Type*⟩ ⟨*Id*⟩ ⟨*Subscripts*⟩ = ⟨*EltInit*⟩ ;
⟨*DisExpr*⟩   →   ⟨*Expr*⟩
            →   < ⟨*Expr*⟩$^+$ >

Figure 8.1: The Syntax of the Display Instructions.

```
enum Cities ...;
enum Products ...;
float+ limit = ...;
float+ supply[Products,Cities] = ...;
float+ demand[Products,Cities] = ...;
assert
   forall(p in Products)
      sum(o in Cities) supply[p,o] = sum(d in Cities) demand[p,d];
float+ cost[Products,Cities,Cities] = ...;

var float+ trans[Products,Cities,Cities];
minimize
   sum(p in Products & o,d in Cities) cost[p,o,d] * trans[p,o,d]
subject to {
   forall(p in Products & o in Cities)
      sum(d in Cities) trans[p,o,d] = supply[p,o];
   forall(p in Products & d in Cities)
      sum(o in Cities) trans[p,o,d] = demand[p,d];
   forall(o, d in Cities)
      sum(p in Products) trans[p,o,d] <= limit;
};

display trans;
```

Statement 8.1: A Transportation Model with a Display Instruction (`transp1.mod`).

8.2 Filtering and Aggregating Results

It is sometimes interesting to filter the results to display the most important information only. In the transportation problem, most variables are zero, so that showing only variables with strictly positive values yields a much more compact display. The display instruction

```
display(o in Orig, d in Dest : trans[o,d] > 0) trans[o,d];
```

implements this idea and produces results of the form

```
Optimal Solution at Cost:   206400.0000

  trans[Bruxelles,Amsterdam] = 600.0000
  trans[Bruxelles,London] = 1100.0000

  trans[Louvain,Lille] = 1200.0000
  trans[Louvain,Nancy] = 600.0000
  trans[Louvain,Koeln] = 400.0000
  trans[Louvain,Calais] = 600.0000

  trans[Namur,Paris] = 1500.0000
  trans[Namur,Amsterdam] = 1100.0000
  trans[Namur,Calais] = 400.0000
```

which captures the relevant information only. Display instructions may use the full expressive power of the formal parameters of OPL and the expression is displayed only for the valid values of these parameters.

It is also useful in some models to aggregate the results to produce a more global view of the solution. For instance, in the transportation problem, it may be relevant to display all shipments from a given city. The instruction

```
display(o in Orig) sum(d in Dest) trans[o,d];
```

displays exactly this information and would produce a result of the form

```
Optimal Solution at Cost:   206400.0000

  sum(d in Dest) trans[Bruxelles,d] = 1700.0000
  sum(d in Dest) trans[Louvain,d] = 2800.0000
  sum(d in Dest) trans[Namur,d] = 3000.0000
```

8.3 Computing Derived Results

The display instructions can in fact be interleaved with data declarations, as in

```
float+ transOrig[o in Orig] = sum(d in Dest) trans[o,d];
display transOrig;
```

These instructions produce essentially the same result as the instruction

```
display(o in Orig) sum(d in Dest) trans[o,d];
```

presented earlier. There are only syntactic differences in the result, which now becomes

```
Optimal Solution at Cost:   206400.0000

  transOrig[Bruxelles] = 1700.0000
  transOrig[Louvain] = 2800.0000
  transOrig[Namur] = 3000.0000
```

These data declarations can also be browsed inside OPL STUDIO and are thus interesting in their own right.

8.4 Displaying Tuples

Another interesting feature of OPL is its ability to display tuples. For instance, the display instruction

```
display(d in Dest) <trans[Namur,d],trans[Namur,d].rc>;
```

displays the transportation variables and the corresponding reduced costs of the variables whose origin is Namur. The result of this display expression is as follows:

```
Optimal Solution at Cost:   206400.0000

  <trans[Namur,Paris],trans[Namur,Paris].rc> = <1500.0000,0.0000>
  <trans[Namur,Lille],trans[Namur,Lille].rc> = <0.0000,2.0000>
  <trans[Namur,Nancy],trans[Namur,Nancy].rc> = <0.0000,2.0000>
  <trans[Namur,Koeln],trans[Namur,Koeln].rc> = <0.0000,1.0000>
  <trans[Namur,Amsterdam],trans[Namur,Amsterdam].rc> = <1100.0000,0.0000>
  <trans[Namur,London],trans[Namur,London].rc> = <0.0000,5.0000>
  <trans[Namur,Calais],trans[Namur,Calais].rc> = <400.0000,0.0000>
```

II THE APPLICATION AREAS

9 Linear and Integer Programming

This chapter studies the application of OPL to linear programming, integer programming, mixed-integer linear programming, and piecewise-linear programming.

9.1 Linear Programming

Linear programming consists of optimizing a linear function subject to linear constraints over real variables. As mentioned earlier, large instances of linear programs can be solved efficiently. This section reviews how to solve linear programs in OPL.

9.1.1 A Production Problem

Consider again the production planning problem of Section 2.1.6. The model is depicted again in Figure 9.1 and the instance data is shown in Figure 9.2. The model aims at minimizing the production cost for a number of products while satisfying customer demand. Each product can be produced either inside the company or outside, at a higher cost. The inside production is constrained by the company's resources, while outside production is considered unlimited. The model first declares the products and the resources. The data consists of the description of the products, i.e., the demand, the inside and outside costs, and the resource consumption, and the capacity of the various resources. The variables for this problem are the inside and outside production for each product. For these statements, OPL returns the optimal solution

```
Optimal Solution with Objective Value 372.0000
   inside[kluski] = 40.0000
   inside[capellini] = 0.0000
   inside[fettucine] = 0.0000

   outside[kluski] = 60.0000
   outside[capellini] = 200.0000
   outside[fettucine] = 300.0000
```

9.1.2 A Multi-Period Production Planning Problem

Large linear-programming problems are often obtained from simpler ones by generalizing them along one or more dimensions. A typical extension of production-planning problems is to consider several production periods and to include inventories in the model. This section presents a multi-period production planning model that generalizes the model of the previous section

```
enum Products ...;
enum Resources ...;
float+ consumption[Products,Resources] = ...;
float+ capacity[Resources] = ...;
float+ demand[Products] = ...;
float+ insideCost[Products] = ...;
float+ outsideCost[Products] = ...;

var float+ inside[Products];
var float+ outside[Products];

minimize
   sum(p in Products) (insideCost[p]*inside[p] + outsideCost[p]*outside[p])
subject to {
   forall(r in Resources)
      sum(p in Products) consumption[p,r] * inside[p] <= capacity[r];
   forall(p in Products)
      inside[p] + outside[p] >= demand[p];
};
```

Statement 9.1: A Production-Planning Problem (`production.mod`).

```
Products = { kluski capellini fettucine };
Resources = { flour eggs };
consumption = [[0.5 0.2] [0.4 0.4] [0.3 0.6]];
capacity = [20, 40];
demand = [100, 200, 300];
insideCost = [0.6, 0.8, 0.3];
outsideCost = [0.8, 0.9, 0.4];
```

Statement 9.2: Instance Data for Production-Planning Problem (`production.dat`).

9. Linear and Integer Programming

The main generalization is to consider the demand for the products over several periods and to allow the company to produce more than the demand in a given period. Of course, there is an inventory cost associated with storing the additional production. Statement 9.3 depicts the new model and Statement 9.4 describes the instance data. Most of the model generalizes smoothly. For instance, the capacity constraints stated for all resources and all periods become

```
forall(r in Resources, t in Periods)
    sum(p in Products) consumption[r,p] * inside[p,t] <= capacity[r];
```

The most novel part of the statement is the constraint linking the demand, the inventory, and the production:

```
forall(p in Products, t in Periods)
    inv[p,t-1] + inside[p,t] + outside[p,t] = demand[p,t] + inv[p,t];
```

The constraint states that, for each product p and each period t, the inventory of period t-1 added to the production of period t is equated to the demand of period t plus the inventory of period t. Of course, the fact that the variables inv[p,t] are constrained to be nonnegative is critical to satisfying the demand and to disallow backorders. The objective function is also generalized to add the inventory costs.

Note also the type declaration

```
struct Plan {
    float+ inside;
    float+ outside;
    float+ inv;
};
```

and the display instructions

```
Plan plan[p in Products, t in Periods] = <inside[p,t],outside[p,t],inv[p,t]>;
display plan;
```

which were added to produce a visually pleasing display. For example, on the instance data depicted in Figure 9.4, OPL produces the optimal solution

```
Optimal Solution with Objective Value:   344.6667

  plan[kluski,1]   = <inside:0.0000,outside:120.0000,inv:110.0000>
  plan[kluski,2]   = <inside:0.0000,outside:0.0000,inv:10.0000>
  plan[kluski,3]   = <inside:10.0000,outside:0.0000,inv:0.0000>
  plan[capellini,1] = <inside:0.0000,outside:320.0000,inv:300.0000>
```

```
plan[capellini,2] = <inside:0.0000,outside:0.0000,inv:100.0000>
plan[capellini,3] = <inside:0.0000,outside:0.0000,inv:0.0000>
plan[fettucine,1] = <inside:66.6667,outside:116.6667,inv:133.3333>
plan[fettucine,2] = <inside:66.6667,outside:0.0000,inv:100.0000>
plan[fettucine,3] = <inside:0.0000,outside:0.0000,inv:0.0000>
```

9.1.3 A Blending Problem

Blending problems are another typical application of linear programming. Consider the following problem. An oil company manufactures three types of gasoline: `super`, `regular`, and `diesel`. Each type of gasoline is produced by blending three types of crude oil: `crude1`, `crude2`, and `crude3`. Figure 9.1 depicts the sales price and the purchase price per barrel of the various products. The gasolines must satisfy some quality criteria with respect to their lead content and their octane ratings, thus constraining the possible blendings. Figure 9.2 describes the relevant instance data. The company must also satisfy its customer demand, which is 3,000 barrels a day of `super`, 2,000 of `regular`, and 1,000 of `diesel`. The company can purchase 5,000 barrels of each type of crude oil per day and can process at most 14,000 barrels a day. In addition, the company has the option of advertising a gasoline, in which case the demand for this type of gasoline increases by ten barrels for every dollar spent. Finally, it costs four dollars to transform a barrel of oil into a barrel of gasoline.

The model is depicted in Statement 9.5 and the instance data is shown in Statement 9.6. The model uses two sets of variables. Variable `a[g]` represents the amount spent in advertising gasoline `g`. Variable `blend[o,g]` represents the number of barrels of crude oil `o` used to produced gasoline `g`. The demand constraints

```
forall(g in Gasolines)
    sum(o in Oils) blend[o,g] = gas[g].demand + 10*a[g];
```

use both types of variables, since `sum(o in Crudes) blend[o,g]` represents the amount of gasoline `g` produced daily. The constraints

```
forall(o in Oils)
    sum(g in Gasolines) blend[o,g] <= oil[o].capacity;
```

capture the purchase limitations for each type of oil. The constraint

```
sum(o in Oils, g in Gasolines) blend[o,g] <= maxProduction;
```

enforces the capacity limitation on production. The constraints

```
forall(g in Gasolines)
```

```
enum Products ...;
enum Resources ...;
int nbPeriods = ...;
range Periods 1..nbPeriods;
struct Plan {
   float+ inside;
   float+ outside;
   float+ inv;
};
float+ consumption[Resources,Products] = ...;
float+ capacity[Resources] = ...;
float+ demand[Products,Periods] = ...;
float+ inCost[Products] = ...;
float+ outCost[Products] = ...;
float+ inventory[Products] = ...;
float+ invCost[Products] = ...;

var float+ inside[Products,Periods];
var float+ outside[Products,Periods];
var float+ inv[Products,0..nbPeriods];

minimize
   sum(p in Products, t in Periods)
      (inCost[p]*inside[p,t] + outCost[p]*outside[p,t] - invCost[p]*inv[p,t])
subject to {
   forall(r in Resources, t in Periods)
      sum(p in Products) consumption[r,p] * inside[p,t] <= capacity[r];
   forall(p in Products, t in Periods)
      inv[p,t-1] + inside[p,t] + outside[p,t] = demand[p,t] + inv[p,t];
   forall(p in Products)
      inv[p,0] = inventory[p];
};
Plan plan[p in Products, t in Periods] = <inside[p,t],outside[p,t],inv[p,t]>;
display plan;
```

Statement 9.3: A Multi-Period Production-Planning Problem (mulprod.mod).

```
Products = { kluski capellini fettucine };
Resources = { flour eggs };
nbPeriods = 3;
consumption =
   [
      [0.5, 0.4, 0.3],
      [0.2, 0.4, 0.6]
   ];
capacity = [20, 40];
demand =
   [
      [10 100 50]
      [20 200 100]
      [50 100 100]
   ];
inventory = [0 0 0];
invCost = [0.1 0.2 0.1];
insideCost = [0.4, 0.6, 0.1];
outsideCost = [0.8, 0.9, 0.4];
```

Statement 9.4: Instance Data for Multi-Period Production-Planning Problem (`mulprod.dat`).

	Sales Price		Purchase Price
super	$ 70	crude1	$ 45
regular	$ 60	crude2	$ 35
diesel	$ 50	crude3	$ 25

Figure 9.1: Prices for the Blending Problem.

	Octane Rating	Lead Content		Octane Rating	Lead Content
super	≥ 10	≤ 1	crude1	12	0.5
regular	≥ 8	≤ 2	crude2	6	2.0
diesel	≥ 6	≤ 1	crude3	8	3.0

Figure 9.2: Octane and Lead Data for the Blending Problem.

9. Linear and Integer Programming

```
enum Gasolines ...;
enum Oils ...;
struct GasType { float+ demand; float+ price; float+ octane; float+ lead; };
struct OilType { float+ capacity; float+ price; float+ octane; float+ lead; };

GasType gas[Gasolines] = ...;
OilType oil[Oils] = ...;
float+ maxProduction = ...;
float+ prodCost = ...;

var float+ a[Gasolines];
var float+ blend[Oils,Gasolines];

maximize
   sum(g in Gasolines, o in Oils)
      (gas[g].price - oil[o].price - prodCost) * blend[o,g]
   - sum(g in Gasolines) a[g]
subject to {
   forall(g in Gasolines)
      sum(o in Oils) blend[o,g] = gas[g].demand + 10*a[g];
   forall(o in Oils)
      sum(g in Gasolines) blend[o,g] <= oil[o].capacity;

   sum(o in Oils, g in Gasolines) blend[o,g] <= maxProduction;

   forall(g in Gasolines)
      sum(o in Oils) (oil[o].octane - gas[g].octane) * blend[o,g] >= 0;
   forall(g in Gasolines)
      sum(o in Oils) (oil[o].lead - gas[g].lead) * blend[o,g] <= 0;
};
```

Statement 9.5: An Oil-Blending Planning Problem (oil.mod).

```
Gasolines = { super regular diesel };
Oils = { crude1 crude2 crude3 };
gas = [
     #<demand:3000 price:70 octane:10 lead:1>#
     #<demand:2000 price:60 octane:8 lead:2>#
     #<demand:1000 price:50 octane:6 lead:1>#];
oil = [
     #<capacity:5000 price:45 octane:12 lead:0.5>#
     #<capacity:5000 price:35 octane:6 lead:2>#
     #<capacity:5000 price:25 octane:8 lead:3>#];
maxProduction = 14000;
prodCost = 4;
```

Statement 9.6: Data for the Oil-Blending Planning Problem (oil.dat).

```
    sum(o in Oils) (oil[o].octane - gas[g].octane) * blend[o,g] >= 0;
forall(g in Gasolines)
    sum(o in Oils) (oil[o].lead - gas[g].lead) * blend[o,g] <= 0;
```

enforce the quality criteria for the gasolines. The objective function

```
sum(g in Gasolines, o in Oils)
    (gas[g].price - oil[o].price - prodCost) * blend[o,g]
- sum(g in Gasolines) a[g]
```

has four parts: the sales price of the gasolines, the purchase cost of the crude oils, the production costs, and the adverting costs. The optimal solution for this problem is

```
Optimal Solution with Objective Value:   287750.0000

  a[super] = 0.0000
  a[regular] = 750.0000
  a[diesel] = 0.0000

  blend[Crude1,super] = 2000.0000
  blend[Crude1,regular] = 2200.0000
```

9. Linear and Integer Programming

```
enum Cities ...;
enum Products ...;
float+ limit = ...;
float+ supply[Products,Cities] = ...;
float+ demand[Products,Cities] = ...;
assert forall(p in Products)
   sum(o in Cities) supply[p,o] = sum(d in Cities) demand[p,d];
float+ cost[Products,Cities,Cities] = ...;
var float+ trans[Products,Cities,Cities];
minimize
   sum(p in Products & o,d in Cities) cost[p,o,d] * trans[p,o,d]
subject to {
   forall(p in Products & o in Cities)
      sum(d in Cities) trans[p,o,d] = supply[p,o];
   forall(p in Products & d in Cities)
      sum(o in Cities) trans[p,o,d] = demand[p,d];
   forall(o, d in Cities)
      sum(p in Products) trans[p,o,d] <= limit;
};
```

Statement 9.7: A Multi-Product Transportation Model (transp1.mod).

```
blend[Crude1,diesel] = 800.0000
blend[Crude2,super] = 1000.0000
blend[Crude2,regular] = 4000.0000
blend[Crude2,diesel] = 0.0000
blend[Crude3,super] = 0.0000
blend[Crude3,regular] = 3300.0000
blend[Crude3,diesel] = 200.0000
```

9.1.4 Exploiting Sparsity

Statement 9.7 gives again the transportation problem presented in Section 6.4. This problem, known as a multicommodity flow problem on a bipartite graph, is a classic transportation problem with the addition of a capacity constraint on the inter-cities connections. The model is, of course, not appropriate for large-scale transportation problems, where only a fraction of the cities are connected and a fraction of the products are sent along the connections. This section discusses how to exploit the sparsity of large-scale problems. OPL offers more support than other modeling languages in this respect, because it can use records and arrays indexed by arbitrary finite sets.

The methodology for exploiting sparsity in OPL consists of mirroring, in the model, the structure of the application. This structure can be inferred from the objective function and the constraints of the application. For instance, the capacity constraint for the transportation application can be phrased as

"The products sent along any given connection may not exceed the given capacity."

This constraint helps identify two main concepts in the application. The first is the connection between two cities, which can be represented explicitly by a data type

```
struct Connection { Cities o; Cities d; };
```

to manipulate connections as first-class objects. The second fundamental concept is the transportation of a product along a connection, called a *route* in this section. Once again, this concept can be represented explicitly by a data type

```
struct Route { Connection e; Products p; };
```

to manipulate routes directly. The supply and demand constraints exhibit two other fundamental concepts: product suppliers (i.e., the association of a product and a city supplying it) and product consumers (i.e., the association of a product and a city consuming it). The data types

```
struct Supplier { Products p; Cities o; };
struct Customer { Products p; Cities d; };
```

may be used to represent them.

Once the concepts are identified, an appropriate data representation can be chosen so that the model can generate constraints efficiently. Of course, the user data is not necessarily expressed in this representation, but it is usually easy in OPL to transform the user data into an appropriate representation. A good representation for this application consists of a set of connections, a set of routes, the cost of the routes, and the demand and supply information, e.g.,

```
{Route} Routes = ...;
float+ cost[Routes] = ...;
{Connection} Connections = { c | <c,p> in Routes };
{Supplier} Suppliers = { <p,c.o> | <c,p> in Routes };
float+ supply[Suppliers] = ...;
{Customer} Customers = { <p,c.d> | <c,p> in Routes };
float+ demand[Customers] = ...;
```

9. Linear and Integer Programming

```
enum Cities ...;
enum Products ...;

struct Connection { Cities o; Cities d; };
struct Route { Connection e; Products p; };
struct Supplier { Products p; Cities o; };
struct Customer { Products p; Cities d; };

{Route} Routes = ...;
{Connection} Connections = { c | <c,p> in Routes };
{Supplier} Suppliers = { <p,c.o> | <c,p> in Routes };
float+ supply[Suppliers] = ...;
{Customer} Customers = { <p,c.d> | <c,p> in Routes };
float+ demand[Customers] = ...;
float+ lim = ...;
float+ cost[Routes] = ...;
{Cities} orig[p in Products] = { c.o | <c,p> in Routes };
{Cities} dest[p in Products] = { c.d | <c,p> in Routes };

assert forall(p in Products)
   sum(o in orig[p]) supply[<p,o>] = sum(d in dest[p]) demand[<p,d>];

var float+ trans[Routes];

minimize
   sum(r in Routes) cost[r] * trans[r]
subject to {
   forall(p in Products & o in orig[p])
      sum(d in dest[p]) trans[<<o,d>,p>] = supply[<p,o>];
   forall(p in Products & d in dest[p])
      sum(o in orig[p]) trans[<<o,d>,p>] = demand[<p,d>];
   forall(c in Connections)
      sum(<c,p> in Routes) trans[<c,p>] <= lim;
};
```

Statement 9.8: A Sparse Multi-Product Transportation Model (transp3.mod).

Note that the connections, suppliers, and customers are derived automatically from the routes. It is also useful to derive the following data to simplify the constraint statement:

```
{Cities} orig[p in Products] = { c.o | <c,p> in Routes };
{Cities} dest[p in Products] = { c.d | <c,p> in Routes };
```

The objective function and the constraints can now be stated naturally. The objective function

> "minimize the transportation costs along all routes"

is expressed elegantly as

```
minimize
    sum(r in Routes) cost[r] * trans[r]
```

The supply constraint, which can be phrased as

> "for every product and every city shipping the product, the summation of all transportations from that city to a city where the product is in demand is equal to the supply of the product at the supplying city"

is formalized by

```
forall(p in Products & o in orig[p])
    sum(d in dest[p]) trans[<<o,d>,p>] = supply[<p,o>];
```

The demand constraints are stated in a similar way. The capacity constraints are stated elegantly as

```
forall(c in Connections)
    sum(<c,p> in Routes) trans[<<c.o,c.d>,p>] <= lim;
```

This statement is efficient, since OPL retrieves the product from the routes in an efficient way when the connection is known. The complete model is shown in Statement 9.8.

Assume now that part of the user data is given by a relational table that contains tuples of the form <o,d,p,c> indicating that a connection between cities o and d transports product p at cost c. This data can be transformed into the representation used in Statement 9.8. The routes can be obtained as

```
{Route} Routes = { <<o,d>,p> | <p,o,d,c> in TableRoutes };
```

and the costs as

```
range Boolean 0..1;
int nbWorkers = ...;
range Workers 1..nbWorkers;
enum Tasks ...;
{Workers} qualified[Tasks] = ...;
int cost[Workers] = ...;

var Boolean hire[Workers];
minimize
   sum(c in Workers) cost[c]*hire[c]
subject to
   forall(j in Tasks)
      sum(c in qualified[j]) hire[c] >= 1;
{Workers} crew = { c | c in Workers :  hire[c] = 1 };
display crew;
```

Statement 9.9: A Set-Covering Model (covering.mod).

```
initialize {
    forall(<p,o,d,c> in TableRoutes)
       cost[<<o,d>,p>] = c;
};
```

Both of these preprocessings are linear in the size of the table.

9.2 Integer Programming

Integer programming is the class of problems defined as the optimization of a linear function subject to linear constraints over integer variables. Integer programs are, in general, much harder to solve than linear programs and the size of integer programs that can be solved efficiently is much smaller than of linear programs. This section reviews a number of typical integer programs.

9.2.1 Set Covering

Consider selecting workers to build a house. The construction of a house can be divided into a number of tasks, each requiring a number of skills (e.g., plumbing or masonry). A worker may or

```
Tasks = { masonry, carpentry, plumbing, ceiling, electricity, heating, insulation,
    roofing, painting, windows, facade, garden, garage, driveway, moving };
nbWorkers = 32;
qualified = [
      { 1 9 19 22 25 28 31 }
      { 2 12 15 19 21 23 27 29 30 31 32 }
      { 3 10 19 24 26 30 32 }
      { 4 21 25 28 32 }
      { 5 11 16 22 23 27 31 }
      { 6 20 24 26 30 32 }
      { 7 12 17 25 30 31 }
      { 8 17 20 22 23 }
      { 9 13 14 26 29 30 31 }
      { 10 21 25 31 32 }
      { 14 15 18 23 24 27 30 32 }
      { 18 19 22 24 26 29 31 }
      { 11 20 25 28 30 32 }
      { 16 19 23 31 }
      { 9 18 26 28 31 32 }];
cost = [1 1 1 1 1 1 1 2 2 2 2 2 2 3 3 3 3 4 4 4 4 5 5 5 6 6 6 7 8 9];
```

Statement 9.10: Instance Data for the Set-Covering Model (covering.dat).

may not perform a task, depending on skills. In addition, each worker can be hired for a cost that also depends on his qualifications. The problem consists of selecting a set of workers to perform all the tasks, while minimizing the cost. This is known as a set-covering problem. The key idea in modeling a set-covering problem as an integer program is to associate a 0/1 variable with each worker to represent whether the worker is hired. To make sure that all the tasks are performed, it is sufficient to choose at least one worker by task. This constraint can be expressed by a simple linear inequality.

Statement 9.9 describes a set-covering model for this problem and Statement 9.10 shows some instance data. The first instruction in the model declares a range Boolean, a range for the workers and an enumerated type for the tasks. The instruction

```
{Workers} qualified[Tasks] = ...;
```

9. Linear and Integer Programming

	Bonn	Bordeaux	London	Paris	Rome
capacity	1	4	2	1	3
store1	20	24	11	25	30
store2	28	27	82	83	74
store3	74	97	71	96	70
store4	2	55	73	69	61
store5	46	96	59	83	4
store6	42	22	29	67	59
store7	1	5	73	59	56
store8	10	73	13	43	96
store9	93	35	63	85	46
store10	47	65	55	71	95

Figure 9.3: Instance Data for the Warehouse-Location Problem.

declares the workers qualified to perform a given task, i.e., `qualified[t]` is the set of workers able to perform task `t`. The problem variables

```
var Boolean hire[Workers];
```

indicates whether a worker is hired for the project. The constraint

```
forall(t in Tasks)
    sum(w in qualified[t]) hire[w] >= 1
```

make sure that each task is covered by at least on worker. Note also the declaration

```
{workers} crew = {w | w in Workers:  hire[w] = 1};
```

which collects all the hired workers in the set `Crew` to produce a more pleasing representation of the results. OPL returns the solution

```
Optimal Solution at Cost:   14
  crew = {23, 25, 26};
```

for the instance data given above.

9.2.2 Warehouse Location

Warehouse location is another typical integer-programming problem. Consider a company that is considering a number of locations for building warehouses to supply its existing stores. Each possible warehouse has a fixed maintenance cost and a maximum capacity specifying how many stores it can support. In addition, each store can be supplied by only one warehouse and the supply cost to the store differs according to the warehouse selected. The application consists of choosing which warehouses to build and which of them should supply the various stores in order to minimize the total cost, i.e., the sum of the fixed and supply costs. The instance used in this section considers five warehouses and 10 stores. The fixed costs for the warehouses are all identical and equal to 30. Figure 9.3 depicts the transportation costs and the capacity constraints.

The key idea in representing a warehouse-location problem as an integer program consists of using a 0-1 variable for each (warehouse, store) pair to represent whether a warehouse supplies a store. In addition, the model also associates a variable with each warehouse to indicate whether the warehouse is selected. Once these variables are declared, the constraints state that each store must be supplied by a warehouse, that each store can be supplied by only an open warehouse, and that each warehouse cannot deliver more stores than its allowed capacity. The most delicate aspect of the modeling is expressing that a warehouse can supply a store only when it is open. These constraints can be expressed by inequalities of the form

```
supply[w,s] <= open[w]
```

which ensures that when warehouse w is not open, it does not supply store s. This follows from the fact that open[w] = 0 implies supply[w,s] = 0. In fact, these constraint can be combined with the capacity constraints to obtain

```
forall(w in Warehouses, s in Stores)
    sum(s in Stores) supply[s,w] <= capacity[w]*open[w];
```

This formulation implies that a closed warehouse has no capacity.

Statement 9.11 describes an integer program for the warehouse-location problem, and Statement 9.12 depicts some instance data. The statement declares the warehouses and the stores, the fixed cost of the warehouses, and the supply cost of a store for each warehouse. The problem variables

```
var Boolean supply[Stores, Warehouses]
```

represents which warehouses supply the stores, i.e., supply[s,w] is 1 if warehouse w supplies store s and zero otherwise. The objective function

```
minimize
    sum(w in Warehouses) fixedCost * open[w] +
    sum(w in Warehouses, s in Stores) supplyCost[s,w] * supply[s,w]
```

expresses the goal that the model minimizes the fixed cost of the selected warehouses and the supply costs of the stores. The constraint

```
range Boolean 0..1;
int fixed = ...;
enum Warehouses ...;
int nbStores = ...;
range Stores 0..nbStores-1;
int capacity[Warehouses] = ...;
int supplyCost[Stores,Warehouses] = ...;

var Boolean open[Warehouses];
var Boolean supply[Stores,Warehouses];

minimize
   sum(w in Warehouses) fixed * open[w] +
   sum(w in Warehouses, s in Stores) supplyCost[s,w] * supply[s,w]
subject to {
   forall(s in Stores)
      sum(w in Warehouses) supply[s,w] = 1;
   forall(w in Warehouses, s in Stores)
      sum(s in Stores) supply[s,w] <= capacity[w]*open[w];
};

display open;
{Stores} storesof[w in Warehouses] = { s | s in Stores : supply[s,w] };
display storesof;
```

Statement 9.11: A Warehouse-Location Model (`warehouse.mod`).

```
fixed = 30;
nbStores = 10;
Warehouses = {Bonn Bordeaux London Paris Rome };
capacity = [1,4,2,1,3];
supplyCost = [
   [20 24 11 25 30]
   [28 27 82 83 74]
   [74 97 71 96 70]
   [ 2 55 73 69 61]
   [46 96 59 83  4]
   [42 22 29 67 59]
   [ 1  5 73 59 56]
   [10 73 13 43 96]
   [93 35 63 85 46]
   [47 65 55 71 95]];
```

Statement 9.12: Data for the Warehouse-Location Model (`warehouse.dat`).

```
forall(s in Stores)
    sum(w in Warehouses) supply[s,w] = 1
```

states that a store must be supplied by exactly one warehouse. The constraint

```
forall(w in Warehouses)
    sum(s in Stores) supply[s,w] <= capacity[w]*open[w];
```

expresses the capacity constraints for the warehouses and makes sure that a warehouse supplies a store only if the warehouse is open. For the instance data depicted in Figure 9.12, OPL returns the optimal solution

```
Optimal Solution with Objective Value:   383

  open[Bonn] = 1
  open[Bordeaux] = 1
  open[London] = 1
  open[Paris] = 0
  open[Rome] = 1
```

9. Linear and Integer Programming

```
storesof[Bonn]     = { 3 }
storesof[Bordeaux] = { 8, 6, 5, 1 }
storesof[London]   = { 9, 7 }
storesof[Paris]    = { }
storesof[Rome]     = { 4, 2, 0 }
```

9.2.3 Fixed-Charge Problems

Fixed-charge problems are another classic application of integer programs. They resemble some of the production problems seen previously but differ in two respects: the production is an integer value (e.g., a factory must produce bikes or toys), and the factories need to rent (or acquire) some tools to produce some of the products. Consider the following problem. A company manufactures shirts, shorts, and pants. Each product requires a number of hours of labor and a certain amount of cloth, and the company has a limited capacity of both. In addition, all of these products can be manufactured only by renting an appropriate machine. The profit on the products (excluding the cost of renting the equipment) are also known. The company would like to maximize its profit.

Statement 9.13 shows a model for fixed charge problems, while Statement 9.14 gives some instance data. The integer program for this model uses two sets of variables: production variables and rental variables. A production variable `produce[p]` describes the production of product p; a rental variable `rent[m]` is a 0-1 variable representing whether machine m is rented. Most of the constraints are similar to traditional production problems and pose few difficulties. The most delicate aspect of the modeling is expressing that a product can be produced only if its equipment is rented. It is not possible to use the same idea as in the warehouse-location problem: e.g., a constraint

`produce[p] <= rent[m]`

is not correct, since `produce[p]` is not a Boolean variable in this model. One might be tempted to write

`produce[p] <= produce[p] * rent[m]`

but this constraint is not linear. The solution adopted in the model is to use an integer representing the maximal production of any product and write

`produce[p] <= maxProduction * rent[m]`

If machine m is rented, then the constraint is redundant, since `maxProduction` is chosen to be larger than `produce[p]`. Otherwise, the right-hand side is zero and product p cannot be manufactured. Note the data representation in this model. The type

```
enum Machines ...;
enum Products ...;
enum Resources ...;
range Boolean 0..1;

int+ capacity[Resources] = ...;
int MaxProduction = max(r in Resources) capacity[r];
int rentingCost[Machines] = ...;

struct ProductType {
   int profit;
   {Machines} machines;
   int use[Resources];
};
ProductType product[Products] = ...;

var Boolean rent[Machines];
var int produce[Products] in 0..MaxProduction;

maximize
   sum(p in Products) product[p].profit * produce[p] -
   sum(m in Machines) rentingCost[m] * rent[m]
subject to {
   forall(r in Resources)
      sum(p in Products) product[p].use[r] * produce[p] <= capacity[r];
   forall(p in Products, m in product[p].machines)
      produce[p] <= MaxProduction * rent[m];
};
```

Statement 9.13: A Fixed-Charge Model (`fixed.mod`).

```
Machines = { shirtM shortM pantM };
Products = { shirt shorts pants };
Resources = { labor cloth };
capacity = [150 160];
rentingCost = [200 150 100];
product = [
    <6 {shirtM} [3 4]>
    <4 {shortM} [2 3]>
    <7 {pantM} [6 4]>];
```

Statement 9.14: Data for the Fixed-Charge Model (`fixed.dat`).

```
struct ProductType {
    int profit;
    {Machines} machines;
    int use[Resources];
};
```

clusters all data related to a product: its profit, the set of machines it requires, and the way it uses the resources. Note also the constraint

```
forall(p in Products, m in product[p].machines)
    produce[p] <= maxProduction * rent[m];
```

which features a `forall` statement that quantifies over each product and over each machine used by the product.

9.2.4 Search Procedures

Integer programs are generally solved using a branch-and-bound algorithm that uses a linear relaxation at each node of the search tree. OPL provides a number of predifined tools to let users specify their own search procedures. A traditional approach in integer programming is to design a search procedure that splits the domain of a variable in two parts and explores both parts nondeterministically. The splitting point is generally the value of the variable in the linear relaxation. Such a procedure can be implemented as follows:

```
while not isBound(x) do
    let v = simplexValue(x) in
```

```
let l = floor(v) in
let v = ceil(v) in
   try x <= l | x >= v endtry
```

This search procedure is supported directly by the instruction `splitLow`. The instruction `splitUp` is similar, except that the order of the alternatives in the `try` instruction is permuted. The instruction `split` is similar but the order is implementation-dependent. The instructions `branch`, `branchLow`, and `branchUp` generalize the `split` instructions to arrays of variables. For instance, assuming that `x` is a one-dimensional array of integer variables, the instruction `branchLow(x)` can be written as

```
while not isBound(x) do
   select(i in Index :  not isBound(x[i]))
      splitLow(x[i]);
```

These branching instructions can also be applied to arrays of arbitrary dimensions. Of course, all the other constructs described in Chapter 7 can be used as well.

9.3 Mixed Integer-Linear Programming

Mixed integer-linear programs are linear programs in which some variables are required to take integer values, and arise naturally in many applications. The integer variables may come from the nature of the products (e.g., a machine may, or may not, be rented). Mixed integer-linear programs are solved using the same technology as integer programs (or vice-versa). For instance, a branch-and-bound algorithm can exploit the linear relaxation and its branching procedure is applied only to integer variables.

Consider how to upgrade the production-planning problem in Section 2.1.6 to include a fixed charge for the products. Statement 9.15 describes the new model and Statement 9.16 describes the instance data. Note that the model now contains two types of variables: 0-1 variables that represent whether to rent a machine and production variables of type `float+`. The product data is enhanced with a field describing the required machine, while the new constraints are modeled as in the fixed-charge problem in Statement 9.13. On the instance data in Figure 9.16, OPL returns the optimal solution

```
Optimal Solution with Objective Value:   378.3333

  rent[m1] = 0
  rent[m2] = 0
  rent[m3] = 1

  inside[kluski] = 0.0000
```

9. Linear and Integer Programming

```
range Boolean 0..1;
enum Products ...;
enum Resources ...;
enum Machines ...;
float+ maxProduction = ...;
struct TypeProductData {
   float+ demand;
   float+ incost;
   float+ outcost;
   float+ use[Resources];
   Machines machine;
};
TypeProductData product[Products] = ...;
float+ capacity[Resources] = ...;
float+ rentCost[Machines] = ...;

var Boolean rent[Machines];
var float+ inside[Products];
var float+ outside[Products];

minimize
   sum(p in Products) (product[p].incost*inside[p] + product[p].outcost*outside[p]) +
   sum(m in Machines) rentCost[m] * rent[m]
subject to {
   forall(r in Resources)
      sum(p in Products) product[p].use[r] * inside[p] <= capacity[r];

   forall(p in Products)
      inside[p] + outside[p] >= product[p].demand;

   forall(p in Products)
      inside[p] <= maxProduction * rent[product[p].machine];
},
```

Statement 9.15: A Fixed-Charge Production Model (prodmilp.mod).

```
Products = { kluski capellini fettucine };
Resources = { flour eggs };
Machines = { m1 m2 m3 };
rentCost = [20 10 5];
maxProduction = 100000;
product = #[
   kluski    : <100 0.6 0.8 [0.5 0.2] m1>
   capellini : <200 0.8 0.9 [0.4, 0.4] m2>
   fettucine : <300 0.3 0.4 [0.3, 0.6] m3>]#;
capacity = [20, 40];
```

Statement 9.16: Data for the Fixed-Charge Production Model (`prodmilp.dat`.)

```
inside[capellini] = 0.0000
inside[fettucine] = 66.6667

outside[kluski]    = 100.0000
outside[capellini] = 200.0000
outside[fettucine] = 233.3333
```

9.4 Piecewise Linear Programming

This section considers piecewise linear programs, which are also useful in simplifying the models for a variety of applications. Piecewise linear programs are in fact syntactic sugar for linear, integer, or mixed integer-linear programs. In other words, a piecewise linear program can always be transformed into a mixed integer linear program and, sometimes, into a linear program. This last case is particularly interesting from a computational standpoint. This section introduces piecewise linear programs using an inventory application. The company Sailco must determine how many sailboats to produce over four time periods. The demand for the four periods is known (40, 60, 75, 25) and, in addition, an inventory of ten boats is available initially. In each period, Sailco can produce 40 boats at a cost of $400 per boat. Additional boats can be produced at a cost of $450 per boat. The inventory cost is $20 per boat and per period.

Statement 9.17 describes a linear program for this application and Statement 9.18 describes the instance data. The key idea underlying this model is to use two sets of variables for describing the

9. Linear and Integer Programming

```
int+ nbPeriods = ...;
range Periods 1..nbPeriods;

float+ demand[Periods] = ...;
float+ regularCost = ...;
float+ extraCost = ...;

assert regularCost <= extraCost;

float+ capacity = ...;
float+ inventory = ...;
float+ inventoryCost = ...;

var float+ regulBoat[Periods];
var float+ extraBoat[Periods];
var float+ inv[0..nbPeriods];

minimize
   regularCost * (sum(t in Periods) regulBoat[t]) +
   extraCost * (sum(t in Periods) extraBoat[t]) +
   inventoryCost * (sum(t in Periods) inv[t])
subject to {
   inv[0] = inventory;
   forall(t in Periods)
      regulBoat[t] <= capacity;
   forall(t in Periods)
      regulBoat[t] + extraBoat[t] + inv[t-1] = inv[t] + demand[t];
};
```

Statement 9.17: A Simple Inventory Model (sailco.mod).

```
nbPeriods = 4;
demand = [40, 60, 75, 25];
regularCost = 400;
extraCost = 450;
capacity = 40;
inventory = 10;
inventoryCost = 20;
```

Statement 9.18: Data for the Simple Inventory Model (`sailco.dat`).

production: variables `regulBoat[t]` represents the number of boats built at the regular price ($400 in the instance data) for period `t`, while `extraBoat[t]` represents the number of extra boats, i.e., boats built at the higher price. The model also contains inventory variables. Most of the constraints are typical for inventory problems. In addition, the constraint

```
forall(t in Periods)
    regulBoat[t] <= capacity;
```

states that there is a capacity constraint on the regular boats. This constraint could be expressed directly as a bound but this is not of concern since it will disappear in the next model. Note also that all the variables will be given integral values for this application, although they are of type float. This is due to the problem structure, not to chance. The constraint matrix of this problem is totally unimodular, which guarantees that the optimum has only integer values for integer data. See for instance [18] for a discussion of total unimodularity. OPL returns the optimal solution

```
Optimal Solution with Objective Value:   78450.0000

  regulBoat[1] = 40.0000
  regulBoat[2] = 40.0000
  regulBoat[3] = 40.0000
  regulBoat[4] = 25.0000

  extraBoat[1] = 0.0000
  extraBoat[2] = 10.0000
  extraBoat[3] = 35.0000
  extraBoat[4] = 0.0000
```

9. Linear and Integer Programming

```
int+ nbPeriods = ...;
range Periods 1..nbPeriods;
float+ demand[Periods] = ...;
float+ regularCost = ...;
float+ extraCost = ...;
float+ capacity = ...;
float+ inventory = ...;
float+ inventoryCost = ...;

var float+ boat[Periods];
var float+ inv[0..nbPeriods];

minimize
   sum(t in Periods) piecewise{ regularCost -> capacity ; extraCost } boat[t] +
   inventoryCost * (sum(t in Periods) inv[t])
subject to {
   inv[0] = inventory;
   forall(t in Periods)
      boat[t] + inv[t-1] = inv[t] + demand[t];
};
```

Statement 9.19: A Piecewise Linear Model for the Simple Inventory Problem (sailcopw.mod).

```
inv[0] = 10.0000
inv[1] = 10.0000
inv[2] = 0.0000
inv[3] = 0.0000
inv[4] = 0.0000
```

on this problem instance. It is interesting to observe that the model does not preclude producing extra boats even if the production of regular boats does not reach its full capacity. This is not an issue in this model, since the extra boats are more expensive and thus are not produced in an optimal solution. It would become an issue, of course, if the cost of the extra boats is less than the regular price (because of, say, economies of scale). This case is discussed later in this section.

A piecewise linear model for this application is given in Figure 9.19. The data description is similar in this model. What differs from the previous model is the choice of variables, the objective function, and the constraints. There is only one type of production variable in this model and hence there is no distinction between "regular" boats and "extra" boats. In this model, boat[t] represents the total production of boats during period t. Even more interesting is how the objective function is described: it makes it explicit that the cost of building the boats is in fact a piecewise linear function of the production

```
piecewise{ regularCost -> quantity:  extraCost } boat[t]
```

OPL recognizes that this statement is in fact a linear program, applies a transformation to obtain Statement 9.17, and returns the same optimal solution.

Note that all piecewise linear programs are not linear programs. Recall the discussion in the previous paragraph and assume that the cost of extra boats decreases to $350, for instance (because of economies of scale). The transformation would not be correct, because a linear program would tend to use "extra" boats before all the "regular" boats have been built. The transformation must enforce a constraint stipulating that "extra" boats can only be used when all the "regular" boats have been manufactured. The resulting program is a mixed integer-linear program and OPL returns the optimal solution

```
Optimal Solution with Objective Value:   72200.0000

  boat[1] = 90.0000
  boat[2] = 0.0000
  boat[3] = 100.0000
  boat[4] = 0.0000

  inv[0] = 10.0000
  inv[1] = 60.0000
  inv[2] = 0.0000
  inv[3] = 25.0000
  inv[4] = 0.0000
```

This solution is interesting since it uses "extra" boats as much as possible while trying to minimize the use of boats in inventory. As a result, there is no production in the second and fourth periods.

Statement 9.17 can be generalized further to include more pieces. Statement 9.20 depicts such a model. The interesting feature is the objective function

```
minimize
```

```
int+ nbPeriods = ...;
range Periods 1..nbPeriods;
int+ nbPieces = ...;
float+ cost[1..nbPieces] = ...;
float+ breakpoint[1..nbPieces-1] = ...;
float+ demand[Periods] = ...;
float+ inventory = ...;
float+ inventoryCost = ...;

var float+ boat[Periods];
var float+ inv[0..nbPeriods];

minimize
   sum(t in Periods)
      piecewise {
         forall(i in 1..nbPieces-1) cost[i] -> breakpoint[i] ;
         cost[nbPieces]
      } boat[t] +
   inventoryCost * (sum(t in Periods) inv[t])
subject to {
   inv[0] = inventory;
   forall(t in Periods) {
      boat[t] + inv[t-1] = inv[t] + demand[t];
   }
};
```

Statement 9.20: A Generalized Piecewise-Linear Model for the Simple Inventory Problem. (sailcopwg.mod)

```
nbPeriods = 4;
demand = [40, 60, 75, 25];
nbPieces = 2;
cost = [400, 450];
breakpoint = [40];
inventory = 10;
inventoryCost = 20;
```

Statement 9.21: Data for the Generalized Piecewise-Linear Model (`sailcopwg1.dat`).

```
nbPeriods = 4;
demand = [40, 60, 75, 25];
nbPieces = 3;
cost = [300, 400, 450];
breakpoint = [30, 40];
inventory = 10;
inventoryCost = 20;
```

Statement 9.22: Another Instance Data for the Generalized Piecewise-Linear Model (`sailcopwg2.dat`).

```
sum(t in Periods)
   piecewise{
      forall(i in 1..nbPieces-1) cost[i] -> breakpoint[i];
      cost[nbPieces]
   } boat[t] +
inventoryCost * (sum(t in Period) inv[t])
```

which is now generic in the number of pieces. Statement 9.21 describes the same instance data for this model.

Consider now adding a constraint stipulating that the maximum number of boats produced in each period cannot exceed fifty on the instance data depicted in Figure 9.22. This new constraint has a dramatic effect on the model, which is now infeasible. Piecewise linear functions can be used here to understand where the infeasibility comes from. The key insight is to replace the capacity

```
nbPeriods = 4;
demand = [40, 60, 75, 25];
nbPieces = 4;
cost = [300, 400, 450, 100000];
breakpoint = [30, 40, 50] ;
inventory = 10;
inventoryCost = 20;
```

Statement 9.23: Instance Data to Deal with Infeasibility (`sailcopwg3.dat`).

constraint by yet another piece in the piecewise linear function and to associate a huge cost with this new piece. Statement 9.23 depicts the instance data needed to do this and OPL produces the following optimal solution:

```
Optimal Solution with Objective Value:   1560600.0000

   boat[1] = 50.0000
   boat[2] = 50.0000
   boat[3] = 65.0000
   boat[4] = 25.0000

   inv[0] = 10.0000
   inv[1] = 20.0000
   inv[2] = 10.0000
   inv[3] = 0.0000
   inv[4] = 0.0000
```

which indicates clearly where the bottlenecks (i.e., the third period) are located. The result may help Sailco to plan ahead and take appropriate measures.

9.4.1 Complexity Issues

It is important to understand, of course, when a piecewise linear program corresponds to a mixed integer-linear program. Figure 9.4 describes the shapes of functions that can be used in objective functions to produce linear programs: convex piecewise linear functions in minimization problems and concave piecewise linear functions in maximization problems. Of course, summations of such

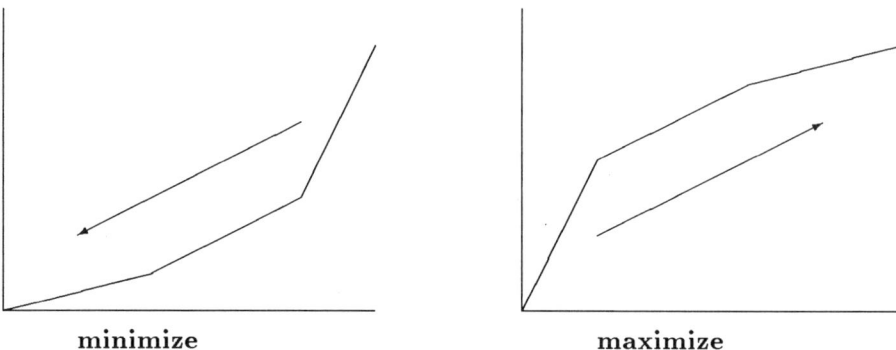

Figure 9.4: Piecewise Linear Functions Leading to Linear Programs.

functions, possibly on different variables, are also appropriate. Similar considerations apply to constraints. A convex piecewise linear function may appear on the left-hand side of an $\leq$ inequality and on the right-hand side of an $\geq$ inequality. A concave piecewise linear function may appear on the right-hand side of an $\leq$ inequality and on the left-hand side of an $\geq$ inequality. In other cases, the piecewise linear program is transformed into a mixed integer-linear program.

9.5 Notes and References

The production problem is adapted from a similar example in [23] and the blending problem is taken from [36]. The discussion of sparsity was inspired by a similar discussion in [9], but the resulting models are different. The warehouse-location instance is taken from [25] and the fixed-charge instance from [36]. The piecewise linear application was also adapted from [36]. The discussion of using piecewise linear functions to analyze infeasibility was motivated by a similar treatment in [9].

10 Constraint Programming

This chapter describes how to use traditional constraint-programming techniques in OPL and illustrates higher-order, global, and nonlinear constraints as well as search procedures. It also discusses how integer programming and constraint programming techniques can be combined in a model.

10.1 Warehouse Location

To illustrate a variety of constraint-programming techniques in OPL, it is useful to reconsider the warehouse-location problem described in Section 9.2.2. In this problem, a company is considering a number of locations for building warehouses to supply its existing stores. Each possible warehouse has a fixed maintenance cost and a maximum capacity specifying how many stores it can support. In addition, each store can be supplied by only one warehouse and the supply cost to the store varies according to the warehouse selected. The application consists of choosing what warehouses to build and which of them should supply the various stores in order to minimize the total cost, i.e., the sum of the fixed and supply costs. The instance used in this section considers five warehouses and 10 stores. The fixed costs for the warehouses are all identical and equal to 30. Figure 10.1 depicts the transportation costs and the capacity constraints.

10.1.1 A Simple Model

Statement 10.1 describes a simple model for the warehouse-location problem and Statement 10.2 shows the instance data again. The data representation is unchanged from the integer-programming formulation. Recall that the basic principle of the integer-programming model for the warehouse location problem consists of using a 0-1 variable for each (warehouse, store) pair to represent whether a warehouse supplies a store. In contrast, the key idea behind the constraint programming model is to use a variable **supplier[s]** to denote the warehouse supplying store s. The model still uses variables of the form **open[w]** to specify whether a warehouse is open. It is not possible to express the problem constraints as linear constraints with this choice of variables, and features from constraint programming are used instead.

The constraints and the objective function are particularly interesting in this model. Consider first the constraint that a store may only be supplied by an open warehouse. It is expressed in the model by the instruction

```
forall(s in Stores)
   open[supplier[s]] = 1;
```

This constraint specifies that warehouse w must be open if it is the supplier of store s. Note that, whenever OPL detects that a warehouse w must be closed in a solution, it automatically removes value w from all the supplier variables. Consider now the capacity constraint which restricts the number of stores which can be supplied by a warehouse. It is expressed by a high-order constraint

	Bonn	Bordeaux	London	Paris	Rome
capacity	1	4	2	1	3
store1	20	24	11	25	30
store2	28	27	82	83	74
store3	74	97	71	96	70
store4	2	55	73	69	61
store5	46	96	59	83	4
store6	42	22	29	67	59
store7	1	5	73	59	56
store8	10	73	13	43	96
store9	93	35	63	85	46
store10	47	65	55	71	95

Figure 10.1: Instance Data for the Warehouse-Location Problem.

```
int fixed = ...;
int nbStores = ...;
enum Warehouses ...;
range Stores 0..nbStores-1;
int capacity[Warehouses] = ...;
int supplyCost[Stores,Warehouses] = ...;

var int open[Warehouses] in 0..1;
var Warehouses supplier[Stores];

minimize
   sum(s in Stores) supplyCost[s,supplier[s]] + sum(w in Warehouses) fixed * open[w]
subject to {
   forall(s in Stores)
      open[supplier[s]] = 1;
   forall(w in Warehouses)
      sum(s in Stores) (supplier[s] = w) <= capacity[w];
};
```

Statement 10.1: A Simple Constraint-Programming Model for Warehouse Location (wlocation.mod).

```
fixed = 30;
nbStores = 10;
Warehouses = {Bonn Bordeaux London Paris Rome };
capacity = [1,4,2,1,3];
supplyCost = [
   [20 24 11 25 30] [28 27 82 83 74] [74 97 71 96 70] [2 55 73 69 61]
   [46 96 59 83 4] [42 22 29 67 59] [1 5 73 59 56] [10 73 13 43 96]
   [93 35 63 85 46] [47 65 55 71 95]];
```

Statement 10.2: Data for the Warehouse-Location Model (`warehouse.dat`).

```
forall(w in Warehouses)
   sum(s in Stores) (supplier[s] = w) <= capacity[w];
```

For each warehouse `w`, the constraint expresses that the number of stores whose supplier is `w` cannot exceed the capacity of `w`. Finally, the objective function

```
minimize
   sum(s in Stores) supplyCost[s,supplier[s]] +
   sum(w in Warehouses) fixed * open[w]
```

expresses the supply costs in an interesting way, since the variables `supplier[s]` are used to index the matrix of costs. The ability to index arrays with variables is fundamental to the constraint programming approach to the model.

10.1.2 A Search Procedure

The model can be enhanced by a search procedure implementing a well-known heuristic called *maximal regret*. To understand the heuristic intuitively, consider what happens when a given store is not assigned to its cheapest supplier. The situation for, say, the second store is not so bad, since its second cheapest supplier (Bonn with cost 28) is roughly equivalent to its cheapest supplier (Bordeaux with cost 27). However, for the fourth store, the situation is more dramatic, since the difference between its cheapest supplier (Bonn with cost 2) and its next cheapest supplier (Bordeaux with cost 55) is very large. The maximal regret heuristic exploits this observation. At any given time of the search, there are a number of stores whose suppliers remain to be determined and a number of warehouses that can be selected. The *regret* of a store at this computation stage is the difference between its first and second choice and the *maximal regret* heuristic recommends

- assign a warehouse to the supplier with the maximal regret;
- try the warehouses in increasing order of cost.

Statement 10.3 depicts a model implementing this heuristic. The model represents the supply costs explicitly to express the heuristic easily. It declares an array

```
var int cost[Stores] in 0..maxCost;
```

to represent these costs and enforces the constraints

```
forall(s in Stores)
    cost[s] = supplyCost[s,supplier[s]];
```

Now the regret of a store can be expressed easily in OPL by the expression

```
regretdmin(cost[s])
```

which is equivalent to

```
dnexthigher(cost[s],dmin(cost[s])) - dmin(cost[s])
```

Using this function, the instruction

```
forall(s in Stores ordered by decreasing regretdmin(cost[s]))
    generate(supplier[s]);
```

is a first approximation to the maximal regret heuristic. The instruction first considers the store with the greatest regret and generates, in a nondeterministic way, a supplier for this store. After constraint solving, it then iterates the process with the remaining stores. The instruction thus implements the heuristic, except that the warehouses are tried in no particular order.

The maximal regret heuristic also recommends trying all the suppliers in increasing order of their costs. To understand how to implement this idea, it is useful to mention that the instruction

```
generate(supplier[s]);
```

is in fact an abbreviation for

```
tryall(w in Warehouses)
    supplier[s] = w;
```

which tries to assign a warehouse w to supplier[s]. The maximal regret heuristic simply specifies the order in which these warehouses must be tried and can be implemented as

```
tryall(w in Warehouses ordered by increasing supplyCost[s,w])
    supplier[s] = w;
```

The complete search strategy is thus implemented as follows:

```
search {
   forall(s in Stores ordered by decreasing regretdmin(cost[s]))
      tryall(w in Warehouses ordered by increasing supplyCost[s,w])
         supplier[s] = w;
   generateSeq(open);
};
```

Note that the instruction generateSeq(open) assigns values to the variables in array open.

10.1.3 An Integrated Integer and Constraint-Programming Model

The integer and constraint models for the warehouse location problem have complementary strengths. Integer programming has the advantage of using linear relaxation to produce a lower bound at each node of the search tree. Constraint programming has the advantage of expressing a heuristic search procedure in an elegant way. These two models can be combined, as shown in Statement 10.4. There are a number of issues to take care of to ensure proper integration. First, note the objective function

```
minimize with linear relaxation
   sum(w in Warehouses) fixed * open[w] +
   sum(w in Warehouses, s in Stores) supplyCost[s,w] * supply[s,w]
```

which uses the keyword with linear relaxation. These keywords are necessary, since the model is no longer a pure integer program (e.g., it contains higher-order constraints). It makes sure that OPL uses the linear relaxation of all linear constraints. The second key aspect is to connect, using constraints, the integer program variables supply and the constraint programming variables suppliers. The constraint

```
forall(s in Stores) supply[s,supplier[s]] = 1;
```

enforces this connection in a simple and elegant way. Finally, the model also connects the cost variables using the instruction

```
forall(s in Stores)
   cost[s] = sum(w in Warehouses) supplyCost[s,w] * supply[s,w];
```

The main point of this section is that combinatorial optimization problems can often be modeled in many ways. OPL makes it possible to combine and integrate these different models in a natural way.

```
int fixed = ...;
int nbStores = ...;
enum Warehouses ...;
range Stores 0..nbStores-1;
int capacity[Warehouses] = ...;
int supplyCost[Stores,Warehouses] = ...;
int maxCost = max(s in Stores & w in Warehouses) supplyCost[s,w];

var int open[Warehouses] in 0..1;
var Warehouses supplier[Stores];
var int cost[Stores] in 0..maxCost;

minimize
   sum(s in Stores) cost[s] + sum(w in Warehouses) fixed * open[w]
subject to {
   forall(s in Stores)
      cost[s] = supplyCost[s,supplier[s]];
   forall(s in Stores)
      open[supplier[s]] = 1;
   forall(w in Warehouses)
      sum(s in Stores) (supplier[s] = w) <= capacity[w];
};

search {
   forall(s in Stores ordered by decreasing regretdmin(cost[s]))
      tryall(w in Warehouses ordered by increasing supplyCost[s,w])
         supplier[s] = w;
   generateSeq(open);
};
```

Statement 10.3: The Maximal Regret Model for the Warehouse-Location Problem (warergt.mod).

10. Constraint Programming

```
range Boolean 0..1;
int fixed = ...;
int nbStores = ...;
enum Warehouses ...;
range Stores 0..nbStores-1;
int capacity[Warehouses] = ...;
int supplyCost[Stores,Warehouses] = ...;
int maxCost = max(s in Stores, w in Warehouses) supplyCost[s,w];

var Boolean open[Warehouses];
var Boolean supply[Stores,Warehouses];
var Warehouses supplier[Stores];
var int cost[Stores] in 0..maxCost;

minimize with linear relaxation
   sum(w in Warehouses) fixed * open[w] +
   sum(w in Warehouses, s in Stores) supplyCost[s,w] * supply[s,w]
subject to {
   forall(s in Stores) sum(w in Warehouses) supply[s,w] = 1;
   forall(w in Warehouses, s in Stores) supply[s,w] <= open[w];
   forall(w in Warehouses) sum(s in Stores) supply[s,w] <= capacity[w]*open[w];

   forall(s in Stores) cost[s] = supplyCost[s,supplier[s]];
   forall(i in Stores) open[supplier[i]] = 1;
   forall(w in Warehouses) sum(s in Stores) (supplier[s] = w) <= capacity[w];

   forall(s in Stores)
      supply[s,supplier[s]] = 1;
   forall(s in Stores)
      cost[s] = sum(w in Warehouses) supplyCost[s,w] * supply[s,w]
};
search {
   forall(s in Stores ordered by decreasing regretdmin(cost[s]))
      tryall(w in Warehouses ordered by increasing supplyCost[s,w])
         supplier[s] = w;
};
```

Statement 10.4: An Integrated Integer and Constraint-Programming Model for Warehouse Location (wareboth.mod).

options	1	2	3	4	5	demand
class 1	y		y	y		1
class 2				y		1
class 3		y			y	2
class 4		y		y		2
class 5	y		y			2
class 6	y	y				2
capacity	1/2	2/3	1/3	2/5	1/5	

Figure 10.2: An Instance of the Car-Sequencing Problem.

10.2 Car Sequencing

The car-sequencing application illustrates several interesting features of constraint programming, including the use of higher-order constraints. It also illustrates the use of redundant (surrogate) constraints to prune the search space more effectively. The problem can be described as follows. Cars in production are placed on an assembly line moving through various units that install options such as air-conditioning and radios. The assembly line can thus be viewed as composed of slots and each car must be allocated to a single slot. The cars cannot be allocated arbitrarily, since the production units have limited capacity and the options must be added to the cars as the assembly line is moving in front of the unit. These *capacity constraints* are formalized using constraints of the form r *outof* s, indicating that, out of each sequence of s cars, the unit can produce at most r cars with the option. The car-sequencing problem amounts to finding an assignment of cars to the slots that satisfies the capacity constraints.

Here is an illustration of the problem on a simple instance. In the instance, and in the model below, cars requiring the same set of options are clustered into classes. Figure 10.2 contains the instance data for a problem with five options, six classes, and 10 cars. Here "y" means that a particular option is required by the class and a blank means that it is not required. The capacity constraint r/s should be read as r *outof* s. For example, two cars of class 6 must be produced, which requires options 1 and 2. The capacity unit for option 1 has a constraint "1 outof 2", indicating that no two consecutive cars can require the option.

The search space in this problem is made up of the possible values for the assembly-line slots. Figures 10.3 and 10.4 depict a solution to the simple instance. Figure 10.3 specifies which class is selected for a given slot, while Figure 10.4 specifies which options are used by the car assigned to a given slot. In the following, we abuse terminology and simply refer to a class of cars as a car.

Statement 10.5 describes a model for the car-sequencing problem and Statement 10.6 gives the instance data. As usual, the first part of the OPL is devoted to the problem data. The model first declares the number of cars, the number of options, and the number of slots, as well as their

10. Constraint Programming

slot	1	2	3	4	5	6	7	8	9	10
class 1	+									
class 2		+								
class 3				+				+		
class 4					+	+				
class 5				+			+		
class 6			+							+

Figure 10.3: A Solution to the Car-Sequencing Instance.

slot	1	2	3	4	5	6	7	8	9	10
option 1	+		+		+			+		+
option 2			+	+		+	+		+	+
option 3	+				+			+		
option 4	+	+				+	+			
option 5				+					+	

Figure 10.4: The Assembly Line in the Solution to the Car-Sequencing Instance.

associated ranges. A structure type is then declared to specify capacity constraints of the form `l outof u`. The next declarations specify the demand for each car (how many of them must be produced), the options required by each car (a two-dimensional Boolean array specifying whether an option is required by a car), and the capacity of the production units for the options. The declaration

```
int optionDemand[i in Options] = sum(j in Cars) demand[j] * option[i,j];
```

defines an array of integers representing the total demand for each option, i.e., it computes the total number of cars requiring each of the options. As shown below, this information is useful in defining redundant constraints that substantially speed up the computation.

The model uses two sets of variables: the *slot* variables, which specify which car is assigned to a given slot in the assembly line, and the *setup* variables, which, given an option and a slot, specify whether the car assigned to the slot requires the option. Obviously, the main output of the model is the slot variables; the setup variables are mostly used to make it easy to state the problem constraints. The possible values for the slot variables are the cars, while the setup variables are simply Boolean variables. For instance, the simple instance declares 10 slot variables taking their values in 1...6 and 50 setup variables.

The first set of constraints expresses the demand constraints, which specify how many cars of each type must be produced. These higher-order constraints state that, for each type of car c, the number of slots assigned to c is equal to the demand for c:

```
int nbCars = ...;
int nbOptions = ...;
int nbSlots = ...;
range Cars 1..nbCars;
range Options 1..nbOptions;
range Slots 1..nbSlots;
int demand[Cars] = ...;
int option[Options,Cars] = ...;
struct Tcapacity { int l; int u; };
Tcapacity capacity[Options] = ...;
int optionDemand[i in Options] = sum(j in Cars) demand[j] * option[i,j];

var Cars slot[Slots];
var int setup[Options,Slots] in 0..1;

solve {
   forall(c in Cars)
      sum(s in Slots) (slot[s] = c) = demand[c];

   forall(o in Options & s in [1..nbSlots - capacity[o].u + 1])
      sum(j in [s ..  s + capacity[o].u - 1]) setup[o,j] <= capacity[o].l;

   forall(o in Options & s in Slots)
      setup[o,s] = option[o,slot[s]];

   forall(o in Options & i in [1..optionDemand[o]])
      sum(s in [1 ..  nbSlots - i * capacity[o].u]) setup[o,s] >=
      optionDemand[o] - i * capacity[o].l;
}
```

Statement 10.5: The Car-Sequencing Problem (car.mod).

10. Constraint Programming

```
nbCars = 6;
nbOptions = 5;
nbSlots = 10;
demand = [1,1,2,2,2,2];
option = [
   [1, 0, 0, 0, 1, 1],
   [0, 0, 1, 1, 0, 1],
   [1, 0, 0, 0, 1, 0],
   [1, 1, 0, 1, 0, 0],
   [0, 0, 1, 0, 0, 0]];
capacity = [<1,2>, <2,3>, <1,3>, <2,5>, <1,5>];
```

Statement 10.6: The Instance Data for the Car-Sequencing Problem (car.dat).

```
forall(c in Cars)
    sum(s in Slots) (slot[s] = c) = demand[c];
```

The next set of constraints is the capacity constraints, which are expressed in terms of the setup variables. The key idea here is as follows. If a capacity constraint for an option o is of the form l outof u, all subsequences of size u of the assembly line are considered and it is necessary that at most l elements of this subsequence require the option. This is expressed in OPL as follows:

```
forall(o in Options & s in [1..nbSlots - capacity[o].u + 1])
    sum(j in [s .. s + capacity[o].u - 1]) setup[o,j] <= capacity[o].l;
```

The instruction considers all options o and (almost) all slots s and makes sure that the number of cars requiring option o on the subsequence starting at s and of size capacity[o].u is not greater than the capacity capacity[o].l. For instance, option 1 (1 outof 2) generates the constraints

```
setup[1,1] + setup[1,2] <= 1;
setup[1,2] + setup[1,3] <= 1;
...
setup[1,9] + setup[1,10] <= 1;
```

while option 2 (2 outof 3) generates the constraints

```
setup[2,1] + setup[2,2] + setup[2,3] <= 2;
setup[2,2] + setup[2,3] + setup[2,4] <= 2;
...
setup[2,8] + setup[2,9] + setup[2,10] <= 2;
```

Although all constraints seem to have been enforced at this point, an important step is still missing. The setup variables and slot variables are not connected: a slot variable can be assigned a value without influencing its corresponding setup variable, and vice versa. To ensure correctness of the results, it is necessary to link the slot and setup variables. The link is easy to establish: given an option o and a slot s, setup[o,s] is 1 if the car assigned to slot s (i.e., slot[s]) requires option o (i.e., option[o,slot[s]] = 1). The resulting set of constraints is specified as follows:

```
forall(o in Options & s in Slots)
    setup[o,s] = option[o,slot[s]];
```

It is clear from the above instruction that the setup variables are not strictly necessary: each variable setup[o,s] could be replaced by option[o,slot[s]]. This would, however, slow down the execution of the model subtantially, since constraint solving over variable-indexed arrays is expensive. It is thus best to factor out these expressions as much as possible.

All constraints have now been defined. However, the efficiency of OPL can be improved by adding redundant (or surrogate) constraints. As mentioned previously, redundant constraints do not remove any solutions: rather, they express properties of the solutions that may help OPL explore the search space more efficiently. In other words, the constraints are semantically redundant but not operationally redundant.

The car-sequencing problem has a redundant constraint worth exploiting. If option o has a capacity constraint l outof u, it follows that the last u slots can contain only l cars, so the other slots must contain all the remaining cars requiring option o, i.e.,

```
setup[o,1] + ...+ setup[o,nbSlots - u] >= optionDemand[o] - l.
```

In the simple instance, option 1 is requested by five cars and has capacity "1 outof 2". Since only one car can be scheduled in the last two slots, four cars must be sequenced in the first eight slots. More generally, the last $k \times r$ slots can only contain $k \times l$ cars, and constraints of the form

```
setup[o,1] + ...+ setup[o,nbSlots - k * u] >= optionDemand[o] - k * l.
```

can be generated for these shorter prefixes. These constraints are stated in OPL as follows:

```
forall(o in Options & i in [1..optionDemand[o]])
    sum(s in [1 ..  nbSlots - i * capacity[o].u]) setup[o,s] >=
    optionDemand[o] - i * capacity[o].l;
```

The effect of these constraints is to prune the search space early and to escape deep backtracking by recognizing and avoiding failures as soon as possible.

It is interesting to study how the model behaves on the simple instance. After the first choice (i.e. slot[1] = 1), the search space and the assembly line are depicted in Figures 10.5 and 10.6.

10. Constraint Programming

slot	1	2	3	4	5	6	7	8	9	10
class 1	+	−	−	−	−	−	−	−	−	−
class 2	−									
class 3	−									
class 4	−									
class 5	−	−	−							
class 6	−	−								

Figure 10.5: Car Sequencing: Search Space After One Choice.

slot	1	2	3	4	5	6	7	8	9	10
option 1	+	−								
option 2	−									
option 3	+	−	−							
option 4	+									
option 5										

Figure 10.6: Car Sequencing: The Assembly Line After One Choice.

slot	1	2	3	4	5	6	7	8	9	10
class 1	+	−	−	−	−	−	−	−	−	−
class 2	−	+	−	−	−	−	−	−	−	−
class 3	−	−								
class 4	−	−	−	−	−					
class 5	−	−	−							
class 6	−	−								

Figure 10.7: Car Sequencing: Search Space After Two Choices (Part I)

slot	1	2	3	4	5	6	7	8	9	10
option 1	+	−								
option 2	−	−								
option 3	+	−	−							
option 4	+	+	−	−	−					
option 5		−								

Figure 10.8: Car Sequencing: The Assembly Line After Two Choices (Part I)

slot	1	2	3	4	5	6	7	8	9	10
class 1	+	−	−	−	−	−	−	−	−	−
class 2	−	+	−	−	−	−	−	−	−	−
class 3	−	−								
class 4	−	−	−	−	−					
class 5	−	−	−	−		−	−		−	−
class 6	−	−								

Figure 10.9: Car Sequencing: The Search Space After Two Choices (Part II)

slot	1	2	3	4	5	6	7	8	9	10
option 1	+	−								
option 2	−	−	+	+	−	+	+	−	+	+
option 3	+	−	−							
option 4	+	+	−	−	−					
option 5		−								

Figure 10.10: Car Sequencing: The Assembly Line After Two Choices (Part II)

slot	1	2	3	4	5	6	7	8	9	10
class 1	+	−	−	−	−	−	−	−	−	−
class 2	−	+	−	−	−	−	−	−	−	−
class 3	−	−			−		−			
class 4	−	−	−	−			−			
class 5	−	−	−	−	+	−	−	+	−	−
class 6	−	−			−		−			

Figure 10.11: Car Sequencing: Search Space After Two Choices (Part III)

slot	1	2	3	4	5	6	7	8	9	10
option 1	+	−		−	+	−	−	+		
option 2	−	−	+	+	−	+	+	−	+	+
option 3	+	−	−		+			+		
option 4	+	+	−	−	−			−		
option 5		−			−			−		

Figure 10.12: Car Sequencing: The Assembly Line After Two Choices (Part III)

10. Constraint Programming

In Figure 10.5, − means that a value is not in the domain of the variable and a + means that a value is assigned to the variable. In Figure 10.6, − means that the value of the variable is 0, while + means that the value of the variable is 1. For instance, the figures show that, since classes 1 and 5 use option 3 whose capacity is 1/3, the second and third slot cannot be given a car of class 5. Now, assigning `slot[2]` with its first possible value, i.e. 2, leads directly to the solution presented at the beginning of the example. Indeed, this choice immediately removes the value 2 for all other slot variables and prevents variables `slot[3]`, `slot[4]`, and `slot[5]` from taking the value 4 because of option 4. This intermediate state is depicted in Figures 10.7 and 10.8. But the redundant constraints for option 2 require that `slot[3]` and `slot[4]` be assigned a class including option 2, since six cars with option 2 must be produced. The effect of these assignments is to fix all option variables concerning option 2 and to remove possible values from the slot variables. The search space at that stage is depicted in Figures 10.9 and 10.10. At this point, the demand constraints for class 5 come into play. Two cars of class 5 must be produced; since only two places are left for them, they are assigned immediately. This leads to the search space depicted in Figures 10.11 and 10.12. The final step amounts to using the redundant constraints for option 1, which fix all option variables related to option 1 and lead to the solution depicted earlier. Note that here the solution was found in two choices without any backtracking.

10.3 The Euler Tour

The Euler tour problem illustrates the global constraint `circuit` and membership constraints. Consider a knight on a chessboard. The problem entails moving the knight according to the rules of chess in order to visit all positions on the chessboard exactly once and return to the initial position. In more technical terms, the problem consists in finding a Hamiltonian circuit in the graph consisting of a node for each position on the chessboard and an edge for each valid knight's move. The problem of finding a Hamiltonian circuit is NP-complete.

Statement 10.7 shows an OPL model for this problem using the numbering convention in Figure 10.13 for the positions. The model associates a variable with each position in the chessboard using the array `jump`, where `jump[i]` is the location to which the knight moves from location i. An interesting feature of this model is the specification of the legal moves of the knight from each position. It uses a generic declaration that computes the values of each of its elements, where the values are sets of integers taken from the range `ChessBoard`. More precisely, `Knightmoves[i]` represents the set of legal moves from position i. In general, for a position i, the moves are given by $i-17, i-15, i-10, i-6, i+6, i+10, i+15$ and $i+17$. For instance, the legal moves from position 37 are 20, 22, 27, 31, 47, 52, and 54. A first attempt to specify `Knightmoves[i]` is as follows:

```
{ChessBoard} Knightmoves[i in ChessBoard] := { j | j in ChessBoard :
        j = i-17 \/ j = i-15 \/ j = i-10 \/ j = i-6
```

```
range ChessBoard 1..64;
{ChessBoard} Knightmove[i in ChessBoard] = { j | j in ChessBoard :
      i mod 8 = 1 &
         (j = i-15 \/ j = i-6 \/ j = i+10 \/ j = i+17)
   \/ i mod 8 = 2 &
         (j=i-17 \/ j=i-15 \/ j=i-6 \/ j=i+10 \/ j=i+15 \/ j=i+17)
   \/ i mod 8 >= 3 & i mod 8 <= 6 &
         (j=i-17 \/ j=i-15 \/ j=i-10 \/ j=i-6 \/ j=i+6 \/ j=i+10 \/ j=i+15 \/ j=i+17)
   \/ i mod 8 = 7 &
         (j=i-17 \/ j=i-15 \/ j=i-10 \/ j=i+6 \/ j=i+15 \/ j=i+17)
   \/ i mod 8 = 0 &
         (j = i-17 \/ j = i-10 \/ j = i+6 \/ j=i+15)
};

var ChessBoard jump[ChessBoard];

solve {
   forall(p in ChessBoard)
      jump[p] in Knightmove[p];

   circuit(jump);

   forall(p in ChessBoard)
      sum(c in Knightmove[p]) (jump[c] = p) = 1;
};
```

Statement 10.7: The Euler Tour Problem (euler.mod).

1	2	3	4	5	6	7	8
9	10	11	12	13	14	15	16
17	18	19	20	21	22	23	24
25	26	27	28	29	30	31	32
33	34	35	36	37	38	39	40
41	42	43	44	45	46	47	48
49	50	51	52	53	54	55	56
57	58	59	60	61	62	63	64

Figure 10.13: The Numbering Used in the Euler Tour.

10. Constraint Programming

```
        \/ j = i+6 \/ j = i+10 \/ j = i+15 \/ j = i+17
};
```

where \/ and & denote "logical or" and "logical and" respectively. It specifies that, for a position i, the set Knightmove[i] is the set of all j in ChessBoard satisfying the equations discussed previously. However, this specification is only correct for columns 3 to 6 and it would generate invalid moves for the other columns: for instance, it would allow a move from position 25 to position 8, which is clearly invalid. It is thus necessary to generalize the above condition slightly through a case analysis, which results in the set definition shown in Statement 10.7. It is interesting to observe the use of arbitrary Boolean combinations of conditions.

There are three types of constraints in the model. The first set of constraints

```
forall(p in ChessBoard)
    jump[p] in Knightmoves[p];
```

are membership constraints that simply specify that variable jump[i] takes only values corresponding to legal uses. The second constraint

```
circuit(jump);
```

specifies that the set of edges

$$\{(1,\text{jump}[1]),\ldots,(64,\text{jump}[64])\}$$

describes a Hamiltonian circuit or, more precisely, that the sequence

$$(1, v_1), (v_1, v_2), \ldots, (v_{n-1}, v_n), (v_n, v_{n+1})$$

where

$$\begin{cases} v_1 = \text{jump}[1] \\ v_i = \text{jump}[v_{i-1}] \quad (2 \leq i \leq 65) \end{cases}$$

is a Hamiltonian circuit. Operationally, circuit(jump) enforces the so-called subtour constraints, i.e., it makes sure that there are no subtours of length less than 64. For instance, if, at some computation stage, jump[1] = 2 and jump[2] = 3, circuit(jump) enforces the two constraints

```
jump[3] <> 2 & jump[3] <> 1
```

The last set of constraints

```
forall(p in ChessBoard)
    sum(c in Knightmoves[p]) (jump[c] = p) = 1;
```

are redundant constraints that make sure that each position in the board is assigned to at least one variable. These constraints are not necessary from a semantic standpoint but they help in pruning the search space.

It is interesting to see at how OPL behaves on this problem. OPL first generates values for the corners, i.e., positions 1, 8, 57, and 64 because these variables have the smallest domains. Note that the assignment

```
jump[1] = 11
```

automatically produces the assignment

```
jump[18] = 1
```

and similarly for the other corners. OPL then continues building these independent paths incrementally, connecting them only later in the computation. In essence, OPL automatically discovers the multi-path method described in Christofides [3].

10.4 Frequency Allocation

The frequency-allocation problem illustrates a number of interesting features of OPL: the use of complex quantifiers, the use of a multi-criterion ordering to choose which variable to assign next, and the functions `abs` and `nbOccur`. It also features an interesting data representation.

The frequency-allocation problem consists of allocating frequencies to a number of transmitters so that there is no interference between transmitters and the number of allocated frequencies is minimized. The problem described here is an actual cellular phone problem where the network is divided into cells, each cell containing a number of transmitters whose locations are specified. The interference constraints are specified as follows:

- The distance between two transmitter frequencies within a cell must not be smaller than 16.
- The distances between two transmitter frequencies from different cells vary according to their geographical situation and are described in a matrix.

The problem of course consists of assigning frequencies to transmitters to avoid interference and, if possible, to minimize the number of frequencies. The rest of this section focuses on finding a solution using a heuristic to reduce the number of allocated frequencies.

Statement 10.8 shows an OPL statement for the frequency-allocation problem and Statement 10.9 describes the instance data. The model data first specifies the number of cells (25 in the instance), the number of available frequencies (256 in the instance), and their associated ranges. The next declarations specify the number of transmitters needed for each cell and the distance between cells. For example, in the instance, cell 1 requires eight transmitters while cell 3 requires six transmitters. The distance between cell 1 and cell 2 is 1.

10. Constraint Programming

```
int nbCells = ...;
int nbFreqs = ...;
range Cells 1..nbCells;
range Freqs 1..nbFreqs;
int nbTrans[Cells] = ...;
int distance[Cells,Cells] = ...;

struct TransmitterType { Cells c; int t; };
{TransmitterType} Transmitters = { <c,t> | c in Cells & t in 1..nbTrans[c] };
var Freqs freq[Transmitters];

solve {
   forall(c in Cells & ordered t1, t2 in 1..nbTrans[c])
      abs(freq[<c,t1>] - freq[<c,t2>]) >= 16;

   forall(ordered c1, c2 in Cells : distance[c1,c2] > 0)
      forall(t1 in 1..nbTrans[c1] & t2 in 1..nbTrans[c2])
         abs(freq[<c1,t1>] - freq[<c2,t2>]) >= distance[c1,c2];
};

search {
   forall(t in Transmitters ordered by increasing <dsize(freq[t]),nbTrans[t.c]>)
      tryall(f in Freqs ordered by decreasing nbOccur(f,freq))
         freq[t] = f;
};
```

Statement 10.8: The Frequency-Allocation Problem (alloc.mod).

```
nbCells = 25;
nbFreqs = 256;
nbTrans = [8 6 6 1 4 4 8 8 8 4 9 8 4 4 10 8 9 8 4 5 4 8 1 1];
distance = [
    [16  1  1  0  0  0  0  0  1  1  1  1  1  2  2  1  1  0  0  0  2  2  1  1  1]
    [1  16  2  0  0  0  0  0  2  2  1  1  1  2  2  1  1  0  0  0  0  0  0  0  0]
    [1   2 16  0  0  0  0  0  2  2  1  1  1  2  2  1  1  0  0  0  0  0  0  0  0]
    [0   0  0 16  2  2  0  0  0  0  0  0  0  0  0  0  1  1  1  0  0  0  1  1  1]
    [0   0  0  2 16  2  0  0  0  0  0  0  0  0  0  0  1  1  1  0  0  0  1  1  1]
    [0   0  0  2  2 16  0  0  0  0  0  0  0  0  0  0  1  1  1  0  0  0  1  1  1]
    [0   0  0  0  0  0 16  2  0  0  1  1  1  0  0  1  1  1  1  2  0  0  0  1  1]
    [0   0  0  0  0  0  2 16  0  0  1  1  1  0  0  1  1  1  1  2  0  0  0  1  1]
    [1   2  2  0  0  0  0  0 16  2  2  2  2  2  2  1  1  1  1  1  1  1  0  1  1]
    [1   2  2  0  0  0  0  0  2 16  2  2  2  2  2  1  1  1  1  1  1  1  0  1  1]
    [1   1  1  0  0  0  1  1  2  2 16  2  2  2  2  2  2  1  1  2  1  1  0  1  1]
    [1   1  1  0  0  0  1  1  2  2  2 16  2  2  2  2  2  1  1  2  1  1  0  1  1]
    [1   1  1  0  0  0  1  1  2  2  2  2 16  2  2  2  2  1  1  2  1  1  0  1  1]
    [2   2  2  0  0  0  0  0  2  2  2  2  2 16  2  1  1  1  1  1  1  1  1  1  1]
    [2   2  2  0  0  0  0  0  2  2  2  2  2  2 16  1  1  1  1  1  1  1  1  1  1]
    [1   1  1  0  0  0  1  1  1  1  2  2  2  1  1 16  2  2  2  1  2  2  1  2  2]
    [1   1  1  0  0  0  1  1  1  1  2  2  2  1  1  2 16  2  2  1  2  2  1  2  2]
    [0   0  0  1  1  1  1  1  1  1  1  1  1  1  1  2  2 16  2  2  1  1  0  2  2]
    [0   0  0  1  1  1  1  1  1  1  1  1  1  1  1  2  2  2 16  2  1  1  0  2  2]
    [0   0  0  1  1  1  2  2  1  1  2  2  2  1  1  1  1  2  2 16  1  1  0  1  1]
    [2   0  0  0  0  0  0  0  1  1  1  1  1  1  1  2  2  1  1  1 16  2  1  2  2]
    [2   0  0  0  0  0  0  0  1  1  1  1  1  1  1  2  2  1  1  1  2 16  1  2  2]
    [1   0  0  0  0  0  0  0  0  0  0  0  0  1  1  1  1  0  0  0  1  1 16  1  1]
    [1   0  0  1  1  1  1  1  1  1  1  1  1  1  1  2  2  2  2  1  2  2  1 16  2]
    [1   0  0  1  1  1  1  1  1  1  1  1  1  1  1  2  2  2  2  1  2  2  1  2 16]];
};
```

Statement 10.9: Instance Data for the Frequency-Allocation Problem (`alloc.dat`).

The first interesting feature of the model is how variables are declared:

```
struct TransmitterType { Cells c; int t; };
{TransmitterType} Transmitters = { <c,t> | c in Cells & t in 1..nbTrans[c] };
var Freqs freq[Transmitters];
```

As is clear from the problem statement, transmitters are contained within cells. The above declarations preserve this structure, which will be useful when stating constraints. A transmitter is simply described as a record containing a cell number and a transmitter number inside the cell. The set of transmitters is computed automatically from the data using

```
{TransmitterType} Transmitters = { <c,t> | c in Cells & t in 1..nbTrans[c] };
```

which considers each cell and each transmitter in the cell. The model then declares an array of variables

```
var Freqs freq[Transmitters];
```

indexed by the set of transmitters; the values of these variables are of course the frequencies associated with the transmitters. There are two main groups of constraints. The first set of constraints handles the distance constraints between transmitters inside a cell. The instruction

```
forall(c in Cells & ordered t1, t2 in 1..nbTrans[c])
   abs(freq[<c,t1>] - freq[<c,t2>]) >= 16;
```

enforces the constraint that the distance between two transmitters inside a cell is at least 16. The instruction is compact mainly because we can quantify several variables in **forall** statements and because of the keyword **ordered**. Note also that the distance is computed using the function **abs**, which computes the absolute value of its argument (which may be an arbitrary integer expression).

The second set of constraints handles the distance constraints between transmitters from different cells. The instruction

```
forall(ordered c1, c2 in Cells : distance[c1,c2] > 0)
   forall(t1 in 1..nbTrans[c1] & t2 in 1..nbTrans[c2])
      abs(freq[<c1,t1>] - freq[<c2,t2>]) >= distance[c1,c2];
```

considers each pair of distinct cells whose distance must be greater than zero and each two transmitters in these cells, and states that the distance between the frequencies of these transmitters must be at least the distance specified in the matrix distance.

Another interesting part of this model is the search strategy. The basic structure is not surprising: OPL considers each transmitter and chooses a frequency nondeterministically. The interesting feature

Rack Model	Power	Connectors	Price
1	150	8	150
2	200	16	200

Figure 10.14: Rack Specifications.

of the model is the heuristic. OPL chooses to generate a value for the transmitter with the smallest domain and, in case of ties, for the transmitter whose cell size is as small as possible. This multi-criterion heuristic is expressed using a tuple <dsize(freq[t]),nbTrans[t.c]> to obtain

```
forall(t in Transmitters ordered by increasing <dsize(freq[t]),nbTrans[t.c]>)
```

Each transmitter is associated with a tuple $< s, c >$, where s is the number of its possible frequencies and c is the number of transmitters in the cell to which the transmitter belongs. A transmitter with tuple $< s_1, c_1 >$ is preferred over a transmitter with tuple $< s_2, c_2 >$ if $s_1 < s_2$ or if $s_1 = s_2$ and $c_1 < c_2$.

Once a transmitter has been selected, OPL generates a frequency for it in a nondeterministic manner. Once again, the model specifies a heuristic for the ordering in which the frequencies must be tried. To reduce the number of frequencies, the model says to try first those values that were used most often in previous assignments. This heuristic is implemented using a tryall instruction with the order specified using the nbOccur function (nbOccur(i,a) denotes the number of occurrences of i in array a at a given step of the execution):

```
forall(t in Transmitters ordered by increasing <dsize(freq[t]),nbTrans[t.c]>)
   tryall(f in Freqs ordered by decreasing nbOccur(f,freq))
      freq[t] = f;
```

On the instance depicted in Statement 10.9, OPL returns a solution with 95 frequencies.

10.5 Rack Configuration

This problem introduces the idea of using constraints to eliminate symmetries and contains some interesting modeling issues. It also illustrates many features found in previous examples, including structures, arrays indexed by variables, and logical connectives. The problem consists of plugging a set of electronic cards into racks with electric connectors. Each card is characterized by the power it requires, while each rack model is characterized by the maximal power it can supply, its number of connectors, and its price. Each card plugged into a rack uses a connector. The purpose of the model is to find an allocation of a given set of cards into the available racks. Figures 10.14 and 10.15 specify an instance for this problem.

10. Constraint Programming

Card Type	Power	Demand
1	20	10
2	40	4
3	50	2
4	75	1

Figure 10.15: Card Specifications.

Statement 10.10 shows an OPL statement for the rack-configuration problem and Statement 10.11 shows the instance display. The model first defines the number of rack models, the number of cards, the number of racks available, and the ranges based on these constants. The model then defines two structure types to describe the properties of rack models and of cards. These types are then used to define the rack models and the cards. Note the dummy rack model that is free but has no power and no connectors (see the instance data in Statement 10.11). The inclusion of this dummy rack model makes the OPL statement simpler to write, as will become clear later on. The next declarations compute automatically the maximum number of connectors, maximum price, and maximum cost. The generic declarations build three arrays to contain the data about the powers, the connectors, and the price of each rack model.

The main idea underlying the OPL statement is to consider all the racks and assign a rack model to them. Not all the racks may be needed in an optimal solution, which is why the dummy rack model is important; The assignment of a rack to the dummy model simply indicates that the rack is not needed. The OPL statement contains two main variables: the array `racks`, which specifies the model assigned to each available rack, and the array `counters`, which specifies how many cards of a given type are assigned to a given rack. In addition, a cost variable specifies the cost of a configuration.

There are three primary sets of constraints in the OPL statement. The first two sets handle the capacity constraints of the racks, and the third makes sure that all the cards are produced. The constraints

```
forall(r in Racks)
    sum(c in Cards) counters[r,c] * car[c].power <= powerData[rack[r]];
```

ensure that the total power of the cards assigned to a given rack does not exceed the power of the rack. Note that `rack[r]` is a variable and is used to index an array. The constraints

```
forall(r in Racks)
    sum(c in Cards) counters[r,c] <= connData[rack[r]];
```

ensure that the number of cards assigned to a given rack does not exceed the number of connectors of the rack. The constraints

```
int nbModel = ...;
int nbCard = ...;
int nbRack = ...;
range Models 0..nbModel-1;
range Cards 0..nbCard-1;
range Racks 0..nbRack-1;
range RacksButFirst 1..nbRack-1;
struct modelType { int power; int connectors; int price; };
struct cardType { int power; int quantity; };
modelType model[Models] = ...;
cardType car[Cards] = ...;

int maxConn = max(r in Models) model[r].connectors;
int maxPrice = max(r in Models) model[r].price;
int maxCost = nbCard * maxPrice;
int powerData[i in Models] = model[i].power;
int connData[i in Models] = model[i].connectors;
int priceData[i in Models] = model[i].price;
var Models rack[Racks];
var int counters[Racks,Cards] in 0..nbCard;
var int cost in 0..maxCost;

minimize
   cost
subject to {
   forall(r in Racks)
      sum(c in Cards) counters[r,c] * car[c].power <= powerData[rack[r]];
   forall(r in Racks)
      sum(c in Cards) counters[r,c] <= connData[rack[r]];
   forall(c in Cards)
      sum(r in Racks) counters[r,c] = car[c].quantity;
   cost = sum(r in Racks) priceData[rack[r]];
};
```

Statement 10.10: The Rack-Configuration Problem (`config.mod`).

```
nbModel = 3;
nbCard = 4;
nbRack = 5;
model = [
    <0, 0, 0>,
    <150, 8, 150>,
    <200, 16, 200>];
car = [
    <20, 10>,
    <40, 4>,
    <50, 2>,
    <75, 1>];
```

Statement 10.11: Instance Data for the Rack-Configuration Problem (`config.dat`).

```
forall(c in Cards)
    sum(r in Racks) counters[r,c] = car[c].quantity;
```

guarantee that the right number of each type of card is produced. The last constraint

```
cost = sum(r in Racks) priceData[rack[r]];
```

simply defines the cost as the sum of the prices of all racks.

The statement described so far is a correct specification of the problem, but unfortunately contains many symmetries; in particular, any permutation of the array rack is also a solution. To remove these symmetries, it is possible to state a constraint forcing an ordering on this array. The constraints

```
forall(r in RacksButFirst)
    rack[r-1] >= rack[r];
```

impose a decreasing ordering on the array, removing a large number of symmetries. Other symmetries remain, however, when the same model is chosen several times. Two racks of the same model are equivalent as far as the statement is concerned and a permutation of their card assignments also leads to a new solution. As a consequence, whenever two racks are of the same model, it is appropriate to state that one of them has at least as many cards as the other:

```
forall(r in RacksButFirst)
    rack[r-1] = rack[r] => counters[r-1,0] >= counters[r,0];
```

Note, however, that the added constraints do not remove all the symmetries. The remaining symmetries are harder to preclude in a static way, but could be removed using a more involved search strategy.

10.6 Notes and References

The first constraint-programming solution for the warehouse-location problem appeared in [28, 32] and featured the `element` constraint to index an array with variables. The first model described in this chapter is based on a refinement of that solution described in [25]. The car-sequencing model is based on [7]. The problem was motivated by an article in AI Expert [19] reporting the failure of an expert system to solve the problem and concluding that fifth-generation tools were not appropriate for the problem. The frequency-allocation problem is taken from [25]. The data-modeling facilities of OPL avoid the tedious encoding that plagues the program described there. The rack-configuration problem is also borrowed from [25] and the model is based on the constraint program described there.

11 Scheduling

This chapter applies OPL to some scheduling and resource-allocation applications. It is generally organized around the various types of resources (e.g., unary and discrete resources and reservoirs). Elements of Chapters 4 and 5 are reprinted in various places throughout to make the presentation more self-contained.

11.1 Origin and Horizon

All scheduling concepts used in OPL are defined over a global time interval

[scheduleOrigin,scheduleHorizon)

closed on the left and open on the right, like all time intervals in OPL. OPL has default values for both the origin (0) and the horizon (a large number). However, it is recommended that they be specified them for particular applications, since smaller time intervals make OPL more space- and time-efficient. The global origin and horizon can be specified by using instructions of the form

```
scheduleOrigin = 0;
scheduleHorizon = 364;
```

and these instructions must be defined before the definition of any other scheduling concepts. This is consistent with our convention of requiring that each object be used only after it has been defined.

11.2 Activities

The most fundamental concept in OPL for scheduling applications is probably the *activity*. An activity can be thought of as an object containing three data items, a starting date, a duration, and an ending date, together with the *duration constraint* stating that the ending date is the starting date plus the duration. In many applications, the duration of an activity is known and the activity is declared together with its duration, as in

```
Activity carpentry(10);
```

which declares an activity `carpentry` whose duration is 10. The starting and ending dates of `carpentry` are integer variables taking their values in the global time interval and consistent with the duration constraints. Activities can also be given a variable duration, in which case the task is usually declared with an integer variable representing the duration, as in

```
var int durationCarpontry in 8..10;
Activity carpentry(durationCarpentry);
```

which declares an activity `carpentry` whose duration is between 8 and 10. The duration variable `durationCarpentry` can appear in the problem constraints and the possible values for duration can thus be further constrained. It is also possible to declare an activity without specifying its duration, as in

```
Activity carpentry;
```

in which case the duration is an integer variable ranging over the interval

```
[0, scheduleHorizon − scheduleOrigin]
```

Arrays of activities can be declared in the usual way and the durations can be specified as described previously. For instance, it is traditional to declare an array of activities as follows:

```
Activity tasks[t in 1..10](duration[t]);
```

The statement declares an array of 10 activities whose activities are `duration[1],...,duration[10]`.

Some activities may be breakable: they can start before a break and be resumed after a break. Activities in OPL are assumed to be unbreakable unless specified otherwise. The declaration

```
Activity a breakable;
```

specifies that `a` is a breakable activity. Of course, breakable activities behave very much like other activities: they have a starting date, an ending date, and a duration, and they may require resources like activities. Their only added functionality is the ability to be interrupted by breaks. It is of course possible to declare arrays of breakable activities. For instance, the declaration

```
Activity tasks[t in 1..10](duration[t]) breakable;
```

defines an array of 10 breakable activities. Finally, it may be useful to declare an array that has both standard and breakable activities. This makes easy to produce generic models that are parametrized by the status of each activity. The declaration

```
Activity tasks[t in 1..10](duration[t]) breakable if t in breakableSet;
```

defines an array of 10 activities, some of which can be breakable.

The starting date, the duration, and the ending date of an activity are accessed as the fields of a structure. For instance, a precedence constraint between two activities `a` and `b` can be specified as follows:

```
b.start >= a.end;
```

or, alternatively,

```
b.start >= a.start + a.duration;
```

In fact, OPL also has a specific constraint for expressing precedence constraints, which is recommended since it produces better visualizations of the results. For instance, a precedence constraint between two activities a and b can be specified as follows:

```
a precedes b;
```

11.3 Unary Resources

A unary resource is one that cannot be shared by two activities; i.e., as soon as an activity requires a resource, for a time interval, no other activity can use it during the same time interval. Unary resources can be used to model a variety of applications. A typical example of a unary resource is a machine in a job-shop scheduling application. This section reviews scheduling applications involving unary resources.

11.3.1 Summary of the Concepts

Unary resources are declared in OPL simply as

```
UnaryResource crane;
```

As usual, it is possible to declare arrays of unary resources, as in

```
UnaryResource machines[1..10];
```

which declares an array of 10 machines.

Once unary resources are declared, it is possible to specify which activities require them. To specify in OPL that an activity excavation requires a resource crane during its execution, it is sufficient to write the constraint

```
excavation requires crane
```

At any time in the computation, OPL makes sure that no two activities requiring the same unary resource are scheduled at the same time. OPL also uses these constraints to update the starting and ending dates of the activities.

11.3.2 The Bridge Problem

The bridge problem involves of finding a schedule that minimizes the time needed to build a five-segment bridge. The project contains a set of 46 tasks and a set of constraints among these tasks (see Figure 11.1). Besides the usual precedence constraints, the problem also contains some resource constraints. Most tasks require a resource (e.g., a crane) and tasks requiring the same resource cannot overlap in time. In addition, the following additional constraints must be satisfied:

1. The time between the completion of a particular formwork and the completion of its corresponding concrete foundation is at most 4 days.

2. There are at most 3 days between the end of a particular excavation (or foundation piles) and the beginning of the corresponding formwork.

3. The formworks must start at least 6 days after the beginning of the erection of the temporary housing.

4. The removal of the temporary housing can start two days before the end of the last masonry work.

5. The delivery of the preformed bearers occurs at least 30 days after the beginning of the project.

Statements 11.1, 11.2, and 11.3 describe an OPL model for the bridge problem. Statement 11.1 depicts the data used in the bridge problem. The enumerated type declarations define the set of tasks and the set of resources. The type declarations define two record types to express distance and precedence constraints. The next declarations specify an array `duration` for the durations of the tasks, various sets of distance constraints, an array `res` which associates each resource with a set of tasks using the resource, and the set of precedence constraints. Note that most of this data is initialized offline (see Statement 11.3). Statement 11.2 is the core of the model. The declaration

```
scheduleHorizon = maxDuration;
```

defines the schedule horizon as the summation of all durations, which is certainly an upper bound to the project duration. The declaration

```
Activity a[t in Task](duration[t]);
```

defines the activities. There is one activity for each enumerated value in `Task` and its duration is `duration[t]`. Recall that an activity is composed of three items: a starting date, a duration, and an ending date, with the constraint that the ending date is equal to the starting date plus the duration. The instruction

```
UnaryResource tool[Resource];
```

declares the unary resources. There is one unary resource for each enumerated value in `Resource`. The rest of Statement 11.2 describes the constraints and the objective function. The objective function simply states that the starting date of the last task must be minimized. The constraint

```
forall(t in precedences)
    a[t.before] precedes a[t.after];
```

states the precedence constraints. An equivalent formulation, as far as the model is concerned, is

11. Scheduling

N	Name	Description	Duration	Resource
1	PA	beginning of project	0	-
2	A1	excavation (abutment 1)	4	excavator
3	A2	excavation (pillar 1)	2	excavator
4	A3	excavation (pillar 2)	2	excavator
5	A4	excavation (pillar 3)	2	excavator
6	A5	excavation (pillar 4)	2	excavator
7	A6	excavation (pillar 5)	5	excavator
8	P1	foundation piles 2	20	pile-driver
9	P2	foundation piles 3	13	pile-driver
10	UE	erection of temporary housing	10	-
11	S1	formwork (abutment 1)	8	carpentry
12	S2	formwork (pillar 1)	4	carpentry
13	S3	formwork (pillar 2)	4	carpentry
14	S4	formwork (pillar 3)	4	carpentry
15	S5	formwork (pillar 4)	4	carpentry
16	S6	formwork (abutment 2)	10	carpentry
17	B1	concrete foundation (abutment 1)	1	concrete-mixer
18	B2	concrete foundation (pillar 1)	1	concrete-mixer
19	B3	concrete foundation (pillar 2)	1	concrete-mixer
20	B4	concrete foundation (pillar 3)	1	concrete-mixer
21	B5	concrete foundation (pillar 4)	1	concrete-mixer
22	B6	concrete foundation (abutment 2)	1	concrete-mixer
23	AB1	concrete setting time (abutment 1)	1	-
24	AB2	concrete setting time (pillar 1)	1	-
25	AB3	concrete setting time (pillar 2)	1	-
26	AB4	concrete setting time (pillar 3)	1	-
27	AB5	concrete setting time (pillar 4)	1	-
28	AB6	concrete setting time (abutment 2)	1	-
29	M1	masonry work (abutment 1)	16	bricklaying
30	M2	masonry work (pillar 1)	8	bricklaying
31	M3	masonry work (pillar 2)	8	bricklaying
32	M4	masonry work (pillar 3)	8	bricklaying
33	M5	masonry work (pillar 4)	8	bricklaying
34	M6	masonry work (abutment 2)	20	bricklaying
35	L	delivery of preformed bearers	2	crane
36	T1	positioning (preformed bearer 1)	12	crane
37	T2	positioning (preformed bearer 2)	12	crane
38	T3	positioning (preformed bearer 3)	12	crane
39	T4	positioning (preformed bearer 4)	12	crane
40	T5	positioning (preformed bearer 5)	12	crane
41	UA	removal of temporary housing	10	-
42	V1	filling 1	15	caterpillar
43	V2	filling 2	10	caterpillar
44	K1	cost point 1	0	-
45	K2	cost point 2	0	-
46	PE	end of project	0	-

Figure 11.1: Data for the Bridge Problem.

```
enum Task ...;
enum Resource ...;
struct Distance {
   Task before;
   Task after;
   int dist;
};
struct Precedence {
   Task before;
   Task after;
};
struct Disjunction {
   Task first;
   Task second;
};
int duration[Task] = ...;
{Distance} max_nf = ...;
{Distance} min_sf = ...;
{Distance} min_ef = ...;
{Distance} min_nf = ...;
{Distance} min_af = ...;
{Task} res[Resource] = ...;
{Precedence} precedences = ...;
int maxDuration = sum(t in Task) duration[t];
```

Statement 11.1: The Bridge Problem (Part I) (bridge.mod).

11. Scheduling

```
scheduleHorizon = maxDuration;

Activity a[t in Task](duration[t]);

UnaryResource tool[Resource];

minimize
   a[stop].start
subject to {
   forall(t in precedences)
      a[t.before] precedes a[t.after];

   forall(t in max_nf)
      a[t.before].end + t.dist >= a[t.after].start;

   forall(t in max_ef)
      a[t.before].end + t.dist >= a[t.after].end;

   forall(t in min_af)
      a[t.before].start + t.dist <= a[t.after].start;

   forall(t in min_sf)
      a[t.before].start + t.dist >= a[t.after].end;

   forall(t in min_nf)
      a[t.before].end + t.dist <= a[t.after].start;

   forall(r in Resource)
      forall(t in res[r])
         a[t] requires tool[r]
};
```

Statement 11.2: The Bridge Problem (Part II) (bridge.mod).

```
Task = {
   start a1 a2 a3 a4 a5 a6 p1 p2 ue s1 s2 s3 s4 s5 s6 b1 b2 b3 b4 b5 b6 ab1 ab2 ab3
   ab4 ab5 ab6 m1 m2 m3 m4 m5 m6 l t1 t2 t3 t4 t5 ua v1 v2 k1 k2 stop};
Resource = {excavator piledriver carpentry concretemixer bricklaying crane caterpillar};
duration = #[
   first:0 a1:4 a2:2 a3:2 a4:2 a5:2 a6:5 p1:20 p2:13 ue:10
   s1:8 s2:4 s3:4 s4:4 s5:4 s6:10 b1:1 b2:1 b3:1 b4:1 b5:1 b6:1
   ab1:1 ab2:1 ab3:1 ab4:1 ab5:1 ab6:1 m1:16 m2:8 m3:8 m4:8 m5:8
   m6:20 l:2 t1:12 t2:12 t3:12 t4:12 t5:12 ua:10 v1:15 v2:10
   k1:0 k2:0 last:0]#;
max_nf = {
   <first l 30> <a1 s1 3> <a2 s2 3> <a5 s5 3> <a6 s6 3> <p1 s3 3> <p2 s4 3> };
min_sf = { <ua m1 2> <ua m2 2> <ua m3 2> <ua m4 2> <ua m5 2> <ua m6 2> };
max_ef = { <s1 b1 4> <s2 b2 4> <s3 b3 4> <s4 b4 4> <s5 b5 4> <s6 b6 4> };
min_nf = { <first l 30> };
min_af = { <ue s1 6> <ue s2 6> <ue s3 6> <ue s4 6> <ue s5 6> <ue s6 6> };
res = #[
   crane :        {t1 t2 t3 t4 t5}
   bricklaying :  {m1 m2 m3 m4 m5 m6}
   carpentry :    {s1 s2 s3 s4 s5 s6}
   excavator :    {a1 a2 a3 a4 a5 a6}
   piledriver :   {p1 p2}
   concretemixer : {b1 b2 b3 b4 b5 b6}
   caterpillar :  {v1 v2}]#;
precedences = {
   <first a1> <first a2> <first a3> <first a4> <first a5> <first a6> <first ue>
   <a1 s1> <a2 s2> <a5 s5> <a6 s6> <a3 p1> <a4 p2> <p1 s3> <p2 s4> <p1 k1> <p2 k1>
   <s1 b1> <s2 b2> <s3 b3> <s4 b4> <s5 b5> <s6 b6> <b1 ab1> <b2 ab2> <b3 ab3> <b4 ab4>
   <b5 ab5> <b6 ab6> <ab1 m1> <ab2 m2> <ab3 m3> <ab4 m4> <ab5 m5> <ab6 m6> <m1 t1>
   <m2 t1> <m2 t2> <m3 t2> <m3 t3> <m4 t3> <m4 t4> <m5 t4> <m5 t5> <m6 t5> <m1 k2>
   <m2 k2> <m3 k2> <m4 k2> <m5 k2> <m6 k2> <l t1> <l t2> <l t3> <l t4> <l t5>
   <t1 v1> <t5 v2> <t2 last> <t3 last> <t4 last> <v1 last> <v2 last> <ua last>
   <k1 last> <k2 last> };
};
```

Statement 11.3: The Bridge Problem (Part III) (bridge.dat).

```
forall(<before,after> in precedences)
   a[before].start + a[before].duration <= a[after].start;
```

The first formulation is recommended, however, since it produces better visualizations of the results. The distance constraints are similar in nature to precedence constraints and should cause little difficulty. The constraint

```
forall(r in Resource)
   forall(t in res[r])
      a[t] requires tool[r]
```

specifies the resource constraints in a concise way. It considers each resource `r` and each task `t` requiring the resource and states that activity `a[t]` requires unary resource `tool[r]` during its execution. Since the resource is unary, this constraint implies that no other task requiring `tool[r]` can overlap in time with `a[t]`. The optimal solution produced by OPL for the bridge problem is

```
Optimal Solution with Objective Value:   104

   a[start] = [0 -- 0 --> 0]
   a[a1] = [4 -- 4 --> 8]
   a[a2] = [2 -- 2 --> 4]
   a[a3] = [8 -- 2 --> 10]
   a[a4] = [0 -- 2 --> 2]
   a[a5] = [13 -- 2 --> 15]
   a[a6] = [35 -- 5 --> 40]
   ...
```

Each activity is displayed by giving its starting date, its duration, and its ending date. For instance,

`a[a5] = [13 -- 2 --> 15]`

reports that activity `a[a5]` starts at day 13, lasts two days, and ends at day 15.

It is interesting to study the computational model for this problem. Since only unary resources are involved, the default search procedure first ranks all unary resources, i.e., to determine, for each resource, a total ordering for all activities requiring the resource Once they are all are ranked, it is easy to find a solution, since the remaining problem is essentially a PERT problem. As a consequence, the search space explored (implicitly) by OPL on this bridge problem is the set of all possible rankings of all unary resources.

11.3.3 The Bridge Problem with Breaks

When modeling real scheduling applications, it may be important to recognize that there are periods, such as weekends, when no activity can be scheduled. As mentioned in Chapter 5, these periods are called *breaks*: OPL offers several tools for specifying breaks. In addition, activities may or may not be interruptible by breaks. *Breakable* activities were discussed previously in Chapter 4.

Consider the bridge problem again and assume that weekends must be taken into account. In addition, assume that tasks L, UE, UA, as well as tasks concerned with the setting of the concrete are not breakable. Figure 11.4 shows the core of the model for this new version of the problem. For simplicity, the model has not been made generic but readers will have no difficulty in doing so. As can be seen, the changes to the model are minimal. The declaration

```
{Task} TasksNotBreakable = { l, ua, ue, ab1, ab2, ab3, ab4, ab5, ab6 };
```

specifies the tasks that are not breakable. The instruction

```
Activity a[t in Task](duration[t]) breakable if t not in TasksNotBreakable;
```

declares the activities as before but, this time, also specifies whether the task is breakable. The specification of the weekends as breaks

```
forall(r in Resources)
    periodicBreak(resource[r],5,2,7);
```

concludes the necessary modifications. The other parts of the statement remains identical. OPL returns an optimal solution of the form

```
Optimal Solution with Objective Value:   142

  a[start] = [0 -- (0) 0 --> 0]
  a[a1]   = [16 -- (4) 6 --> 22]
  a[a2]   = [4 -- (2) 4 --> 8]
  a[a3]   = [8 -- (2) 2 --> 10]
  a[a4]   = [0 -- (2) 2 --> 2]
  a[a5]   = [2 -- (2) 2 --> 4]
  a[a6]   = [29 -- (5) 7 --> 36]
  ...
```

Consider, for instance, the result

```
a[a1] = [16 -- (4) 6 --> 22]
```

It specifies that activity a[a1] starts at day 16 and ends at day 22. Activity a[a1] has a duration of 4 but, since it is interrupted by a break of duration 2, it is actually scheduled for six days.

11. Scheduling

```
{Task} TasksNotBreakable = { l, ua, ue, ab1, ab2, ab3, ab4, ab5, ab6 };

scheduleHorizon = maxDuration;
Activity a[t in Task](duration[t]) breakable if t not in TasksNotBreakable;

UnaryResource tool[Resource];

minimize
   a[stop].start
subject to {
   forall(r in Resource)
      periodicBreak(tool[r],5,2,7);

   forall(t in precedences)
      a[t.before] precedes a[t.after];

   forall(t in max_nf)
      a[t.before].end + t.dist >= a[t.after].start;

   forall(t in max_ef)
      a[t.before].end + t.dist >= a[t.after].end;

   forall(t in min_af)
      a[t.before].start + t.dist <= a[t.after].start;

   forall(t in min_sf)
      a[t.before].start + t.dist >= a[t.after].end;

   forall(t in min_nf)
      a[t.before].end + t.dist <= a[t.after].start;

   forall(r in Resource)
      forall(t in res[r])
         a[t] requires tool[r]
};
```

Statement 11.4: The Bridge Problem with Breaks (Part II) (`bridgebr.mod`).

11.3.4 Job-Shop Scheduling

Job-shop scheduling problems can be modeled easily in OPL. Statement 11.5 describes a simple job-shop scheduling model. The problem is to schedule a number of jobs on a set of machines to minimize completion time, often called the *makespan*. Each job is a sequence of tasks and each task requires a machine. Statement 11.5 first declares the number of machines, the number of jobs, and the number of tasks in the jobs. The main data of the problem, i.e., the duration of all the tasks and the resources they require, are then given. The instructions

```
Activity task[j in Jobs, t in Tasks](duration[j,t]);
Activity makespan(0);

UnaryResource tool[Machines];
```

declares the activities and the unary resources. Note that the makespan is modeled for simplicity as an activity of duration zero. The first set of constraints specifies that the makespan is not smaller than the ending time of the last task of each job. The next two sets specify the precedence and disjunctive constraints. On the instance data shown in Statement 11.6, OPL returns an optimal solution of the form

```
Optimal Solution with Objective Value:   55

  task[1,1] = [5  -- 1 --> 6]
  task[1,2] = [6  -- 3 --> 9]
  task[1,3] = [16 -- 6 --> 22]
  task[1,4] = [30 -- 7 --> 37]
  task[1,5] = [42 -- 3 --> 45]
  task[1,6] = [49 -- 6 --> 55]
  ...
```

11.3.5 Search Procedures

In scheduling applications with unary resources, the general strategy is to rank each unary resource. Ranking a unary resource consists of finding a total ordering for all activities requiring the resource. Once activities are ordered, a solution can be found efficiently. The constructs and procedures available for search procedures were discussed in Chapter 7. This section reviews some of the tools available to rank unary resources.

The highest-level tool available in OPL to assist the ranking process is a nondeterministic instruction **rank** that ranks all unary resources in the schedule. This instruction is implemented in terms of simpler instructions of the form

11. Scheduling

```
int nbMachines = ...;
range Machines 1..nbMachines;
int nbJobs = ...;
range Jobs 1..nbJobs;
int nbTasks = ...;
range Tasks 1..nbTasks;

Machines resource[Jobs,Tasks] = ...;
int+ duration[Jobs,Tasks] = ...;
int totalDuration = sum(j in Jobs, t in Tasks) duration[j,t];

ScheduleHorizon = totalDuration;
Activity task[j in Jobs, t in Tasks](duration[j,t]);
Activity makespan(0);

UnaryResource tool[Machines];

minimize
   makespan.end
subject to {
   forall(j in Jobs)
      task[j,nbTasks] precedes makespan;

   forall(j in Jobs)
      forall(t in 1..nbTasks-1)
         task[j,t] precedes task[j,t+1];

   forall(j in Jobs)
      forall(t in Tasks)
         task[j,t] requires tool[resource[j,t]];
};
```

Statement 11.5: A Job-Shop Scheduling Model (jobshop.mod).

```
nbMachines = 6;
nbJobs = 6;
nbTasks = 6;

resource = [
   [3 1 2 4 6 5]
   [2 3 5 6 1 4]
   [3 4 6 1 2 5]
   [2 1 3 4 5 6]
   [3 2 5 6 1 4]
   [2 4 6 1 5 3]];
duration = [
   [1 3 6 7 3 6]
   [8 5 10 10 10 4]
   [5 4 8 9 1 7]
   [5 5 5 3 8 9]
   [9 3 5 4 3 1]
   [3 3 9 10 4 1]];
```

Statement 11.6: Data for the Job-Shop Scheduling Model (jobshop.dat).

11. Scheduling

Function	Type	Semantics
isRanked(Unary)	int	1 if the resource is ranked and 0 otherwise
isRanked([Unary])	int	1 if all resources are ranked and 0 otherwise
nbPossibleFirst(Unary)	int	number of activities which can be ranked first
nbPossibleLast(Unary)	int	number of activities which can be ranked last
isPossibleFirst(Unary r,Activity a)	int	true iff a can potentially be ranked first on r
isPossibleLast(Unary,Activity)	int	true iff a can potentially be ranked last on r
localSlack(Unary)	int	local slack of the resource
globalSlack(Unary)	int	global slack of the resource

Figure 11.2: Reflective Functions on Unary resources.

```
rank(u)
```

where `rank(u)` ranks all activities using unary resource u.

Choosing which resource to rank next may be important in some applications. OPL supports this process through a number of reflective functions depicted in Figure 11.2. The *global slack* of a resource considers all activities that are not yet ranked, computes the earliest starting date s, the latest finishing date e, and the total duration d of all these activities, and returns (e - s) - d. This produces an approximation of the tightness of the resource at a given computation point. The local slack of a unary resource, another measure of tightness that is more precise but more expensive to compute, entails computing the global slack for a variety of subsets of the unranked activities. Using these functions, it is possible to define search strategies such as

```
forall(r in Resources ordered by increasing localSlack(tool[r]))
   rank(u[r]);
```

It is interesting to consider how to rank a resource. Ranking a resource u could be handled as follows:

```
while not isRanked(u) do
   select(t in Tasks :  not isRanked(u,a[t]))
      tryRankFirst(u,a[t]);
```

assuming that array a indexed by elements of Tasks contains all the activities requiring resource u. The condition `isRanked(u)` returns true if and only if resource u is ranked. It can also be applied to an array of unary resources, in which case it returns true whenever all resources are ranked. The condition

```
isRanked(u,a)
```

returns true whenever activity a is already ranked on unary resource u. The instruction

```
tryRankFirst(u,a)
```

is nondeterministic and has two alternatives. The first alternative adds the constraint that **a** be ranked first among the unranked activities of **u**. The second alternative (used on backtracking) adds the constraint specifying that **a** cannot rank first among the unranked activities of **u**. OPL also provides a nondeterministic instruction `tryRankLast` that tries to rank an activity last among the unranked activities of a resource.

Once again, it may be important to choose which activity to rank next. OPL supports this process in a variety of ways. First, it is possible to use the starting dates of each activity and predefined functions available on variables. For instance, the heuristic that ranks next the variable with the earliest starting date can be expressed as

```
select(t in Tasks :  not isRanked(u,a[t]) ordered by increasing dmin(a[t].start))
   tryRankFirst(u,a[t]);
```

In addition to that support, OPL provides also two other predefined functions. The function

```
isPossibleFirst(u,a)
```

holds if activity **a** can potentially be ranked first on resource **u**. Function

```
isPossibleLast(u,a)
```

holds if activity **a** can potentially be ranked last on resource **u** at this computation stage.

Although in the above discussion a resource is ranked completely before considering the next resource, other strategies are of course possible. For instance, the excerpt

```
while not isRanked(tool) do
   select(r in Resources :  not isRanked(tool[r]))
      select(t in tasks[r] :  not isRanked(tool[r],a[t])
         tryRankFirst(tool[r],a[t]);
```

selects a resource and an activity and tries to rank first the activity on the resource. It then selects a resource again (which may be different from the resource first selected if heuristics are used for the selection) and an activity.

11.4 Discrete Resources

A discrete resource is a resource with a discrete capacity. The capacity, which may vary over time, represents the number of available copies (or instances) of the resource. For instance, a discrete resource may be used to model a budget, as in Chapter 2, a set of "equivalent" workers, or a set of similar machines. This section presents several applications of discrete resources that complement the problem described in Section 2.3.

11.4.1 Summary of the Concepts

A discrete resource is a resource whose capacity is a strictly positive integer. Discrete resources are declared in OPL by specifying their capacity, as in

```
DiscreteResource crane(3);
```

which specifies that `crane` is a resource of capacity 3, i.e., there are three cranes available. It is possible to declare arrays of discrete resources, as in

```
DiscreteResource res[t in 1..10](cap[t]);
```

which declares an array of 10 resources, the capacity of resource i being `cap[i]`. While a unary resource is a discrete resource of capacity 1, the algorithms for unary resources are optimized to exploit all properties of this special case.

Activities can require discrete resources in the same way as unary resources. In addition, an activity can require a certain capacity of the discrete resource. For instance, the constraint

```
a requires(2) crane;
```

specifies that activity `a` requires two cranes. The capacity requested can be an arbitrary integer expression, possibly containing variables. The underlying algorithms in OPL ensure that the amount requested at any time does not exceed the capacity of the resource. In fact, three levels of pruning can be achieved in OPL for discrete resources: the default level, the *disjunctive* level, and the *edge-finder* level. The *disjunctive* level can be requested by declarations of the form

```
DiscreteResource crane(3) using disjunctive;
```

This level of pruning makes sure that, if the total demand of a set of activities requires more than the capacity of the resource, at least one of these activities is scheduled before another activity in the set. The *edge-finder* level can be requested by declarations of the form

```
DiscreteResource crane(3) using edgeFinder;
```

This *edgeFinder* level generalizes the edge-finding algorithm of unary resources to discrete resources. It tries to deduce which activities must be scheduled first (or last) in a set whose total demand exceeds the capacity of the resource. Which level of propagation is appropriate depends, of course, on the application at hand.

As mentioned previously, a constraint

```
a requires(2) r
```

specifies that activity a requires 2 units of the resource r during its execution. As a consequence, as soon as activity a terminates, its requested capacity is returned to the resource r and is available for other activities. For some applications (e.g., when the discrete resource denotes a budget), the requested capacity should not be returned to the resource: it is consumed. This functionality is obtained in OPL by stating constraints of the form

```
a consumes(2) r;
```

Note that it is also possible to use `consumes` for unary resources, although in general this is not particularly useful.

The capacity of a discrete resource may also vary over time, generalizing the concept of breaks in unary resources. The capacity of a discrete resource over time can be specified with constraints of the form

```
capacityMax(<DiscreteResource>,<Start>,<End>,<Cap>)
capacityMin(<DiscreteResource>,<Start>,<End>,<Cap>)
```

A constraint `capacityMax(d,s,e,c)` specifies that the capacity of discrete resource d required by activities over the interval [s,e] is at most c, while a constraint `capacityMin(d,s,e,c)` specifies that the capacity of discrete resource d required by activities over the interval [s,e] is at least c. Note that this last constraint in fact constrains the scheduling of activities.

11.4.2 The Ship-Loading Problem

This application schedules the loading of a ship. It consists of 34 activities subject to precedence constraints, as depicted in Figure 11.3. In addition, all of these activities have durations and require a unique resource of capacity 8 in various quantities, as depicted in Figure 11.4. For instance, activity 1 has a duration of 3 and requests 4 units of the discrete resource. The goal of the application is to minimize the makespan (i.e., the time to load the ship) while satisfying the precedence and capacity constraints. Figure 11.7 gives an OPL model for solving this problem.

It is interesting to review some of the new functionalities of OPL used in this model. First, the declaration

```
DiscreteResource res(8);
```

declares a discrete resource of capacity 8. Second, the constraint

```
forall(t in Tasks)
    a[t] requires(demand[t]) res;
```

specifies that activity `a[t]` requires `demand[t]` units of the discrete resource. For this application, OPL returns an optimal solution of the form

11. Scheduling

Activity	Successors	Activity	Successors	Activity	Successors	Activity	Successors
1	2 4	11	13	21	22	31	28
2	3	12	13	22	23	32	33
3	5 7	13	15 16	23	24	33	34
4	5	14	15	24	25	34	
5	6	15	18	25	26 30 31 32		
6	8	16	17	26	27		
7	8	17	18	27	28		
8	9	18	19 20 21	28	29		
9	10 14	19	23	29			
10	11 12	20	23	30	28		

Figure 11.3: Precedence Constraints for the Ship-Loading Problem.

Act.	Dur.	Cap.	Act.	Dur.	Cap.	Act.	Dur.	Cap.	Act.	Dur.	Cap.
1	3	4	11	3	4	21	1	4	31	2	3
2	4	4	12	2	5	22	2	4	32	1	3
3	4	3	13	1	4	23	4	7	33	2	3
4	6	4	14	5	3	24	5	8	34	2	3
5	5	5	15	2	3	25	2	8			
6	2	5	16	3	3	26	1	3			
7	3	4	17	2	6	27	1	3			
8	4	3	18	2	7	28	2	6			
9	3	4	19	1	4	29	1	8			
10	2	8	20	1	4	30	3	3			

Figure 11.4: Capacity Constraints for the Ship-Loading Problem.

```
int capacity = 8;
int nbTasks = 34;
range Tasks 1..nbTasks;
int duration[Tasks] = [
   3, 4, 4, 6, 5, 2, 3, 4, 3, 2, 3, 2, 1, 5, 2, 3, 2, 2, 1, 1,
   1, 2, 4, 5, 2, 1, 1, 2, 1, 3, 2, 1, 2, 2];
int totalDuration = sum(t in Tasks) duration[t];
int demand[Tasks] = [
      4, 4, 3, 4, 5, 5, 4, 3, 4, 8, 4, 5, 4, 3, 3, 3, 6, 7, 4, 4,
      4, 4, 7, 8, 8, 3, 3, 6, 8, 3, 3, 3, 3, 3];
struct Precedences {
   int before;
   int after;
};
{Precedences} setOfPrecedences = {
      <1, 2>, <1, 4>, <2, 3>, <3, 5>, <3, 7>, <4, 5>, <5, 6>, <6, 8>, <7, 8>, <8, 9>,
      <9, 10>, <9, 14>, <10, 11>, <10, 12>, <11, 13>, <12, 13>, <13, 15>, <13, 16>,
      <14, 15>, <15, 18>, <16, 17>, <17, 18>, <18, 19>, <18, 20>, <18, 21>, <19, 23>,
      <20, 23>, <21, 22>, <22, 23>, <23, 24>, <24, 25>, <25, 26>, <25, 30>, <25, 31>,
      <25, 32>, <26, 27>, <27, 28>, <28, 29>, <30, 28>, <31, 28>, <32, 33>, <33, 34> };

scheduleHorizon = totalDuration;
Activity a[t in Tasks](duration[t]);
DiscreteResource res(8);
Activity makespan(0);
minimize
   makespan.end
subject to {
   forall(t in Tasks)
      a[t] precedes makespan;
   forall(p in setOfPrecedences)
      a[p.before] precedes a[p.after];
   forall(t in Tasks)
      a[t] requires(demand[t]) res;
};
```

Statement 11.7: The Ship-Loading Problem (shipload.mod).

```
Optimal Solution with Objective Value:  66

    a[1] = [0 -- 3 --> 3]
    a[2] = [3 -- 4 --> 7]
    a[3] = [7 -- 4 --> 11]
    a[4] = [3 -- 6 --> 9]
    ...
```

Observe that, at time 3, activities a[2] and a[4] both require 4 units of the resource, implying that no other activity can be scheduled at this time. The efficiency of the model can be improved by using the choice specification

```
search {
    setTimes
};
```

since the application involves only precedence and resource constraints. It is important to stress here that the solution assigns to an activity the resource necessary at any given time during its execution. However, the resource may be different at two given times, since the resources are considered "equivalent". Section 11.4.4 discusses how to transform discrete resources into a set of unary resources for some applications.

11.4.3 The Perfect Square Problem Revisited

This second application of discrete resources reconsiders the perfect square problem described in Section 2.2.4. Statement 11.8 gives a model for this problem using discrete resources. The key insight is to represent a square by two activities x[i] and y[i] whose durations are the size of the square. Of course, the starting dates of x[i] and y[i] represent the coordinates of the bottom-left corner of the square. The master square is approximated by two discrete resources rx and ry whose capacities are the size of the master square. The activities x[i] require discrete resource rx while the activities y[i] require ry, both equal in capacity to the square they represent (or, as a matter of fact, to their duration): i.e.,

```
forall(s in Squares) {
    x[s] requires(size[s]) rx;
    y[s] requires(size[s]) ry
};
```

In addition, the model imposes capacity constraints on the discrete resources to ensure that there is no empty space, i.e.,

```
capacityMin(rx,0,SizeSquare,SizeSquare);
capacityMin(ry,0,SizeSquare,SizeSquare);
```

Of course, these constraints are only approximations and do not capture the two-dimensionality of the application. They are best viewed as redundant or surrogate constraints. The only "real" constraints are the non-overlapping constraints, i.e.,

```
forall(ordered i, j in Squares)
   x[i].end <= x[j].start \/ x[j].end <= x[i].start \/
   y[i].end <= y[j].start \/ y[j].end <= y[i].start;
```

11.4.4 From Discrete to Unary Resources

For applications using a collection of unary resources that are considered "equivalent", it is often of benefit to use a discrete resource and then convert the solution of the discrete resource problem into a solution to the original problem. This methodology avoids the exploration of a set of highly symmetrical configurations and thus may significantly reduce the search space. For instance, a job shop may have three machines with identical functionalities and the application may not care which of them is actually used for performing a given task. Discrete resources are particularly attractive in this case because, as soon as a solution to the discrete resource problem has been found, it can easily be converted into an actual solution to the problem with unary resources.

This principle can be illustrated on a variant of the house problem discussed in Section 2.3. In this variant, the tasks are still subject to the same precedence constraints but the budget constraints are dropped. Instead, each task requires a worker. There are three workers; `Thomas`, `Maite`, and `Antoine`, who are considered "equivalent" for the purpose of the application. The problem is to minimize the completion date subject to these constraints.

Statement 11.9 gives an `OPL` model for this problem in which each worker is a unary resource. In addition, the model declares a discrete resource `pool` of capacity 3 that aggregates the workers.[1] It also uses a array `use` of 0/1 variables to specify which worker is associated with the tasks. The problem constraints consist mainly of the precedence constraints and the discrete resource constraints

```
forall(t in Tasks)
   a[t] requires(1) pool;
```

The remaining constraints

```
forall(t in Tasks, w in Workers)
```

---

[1] Once again, the model could be made completely generic; it is specialized to the instance data only for simplicity.

11. Scheduling

```
int SizeSquare = 112;
int NbSquares = 21;
range Squares 1..NbSquares;
range Positions 1..SizeSquare;
int size[Squares]=[50,42,37,35,33,29,27,25,24,19,18,17,16,15,11,9,8,7,6,4,2];

scheduleHorizon = SizeSquare;
Activity x[s in Squares](size[s]);
Activity y[s in Squares](size[s]);
DiscreteResource rx(SizeSquare);
DiscreteResource ry(SizeSquare);

solve {
   capacityMin(rx,0,SizeSquare,SizeSquare);
   capacityMin(ry,0,SizeSquare,SizeSquare);
   forall(ordered i, j in Squares)
      x[i].end <= x[j].start \/ x[j].end <= x[i].start \/
      y[i].end <= y[j].start \/ y[j].end <= y[i].start;
   forall(s in Squares) {
      x[s] requires(size[s]) rx;
      y[s] requires(size[s]) ry
   };
};
search {
   setTimes(x);
   setTimes(y);
};
```

Statement 11.8: The Perfect Square Problem Revisited (squarea.mod).

```
a[t] requires(use[t,w]) worker[w];

forall(t in Tasks)
    sum(w in Workers) use[t,w] = 1;
```

are essentially inactive during the search, as will shortly become clear. They specify that each task must be assigned a unique worker. Note that the constraints

```
forall(t in Tasks, w in Workers)
    a[t] requires(use[t,w]) worker[w];
```

require capacity zero or one, depending on the value of the variables in **use**. This functionality is discussed again later in this chapter.

The search procedure here is most interesting. The instruction `setTimes(a);` actually solves the problem: it assigns a starting date to each activity so that the discrete resource constraints are satisfied. When this instruction succeeds, the existence of a solution is guaranteed. The remaining part of the search procedure

```
once {
    forall(t in Tasks ordered by increasing dmin(a[t].start))
        tryall(w in Workers)
            use[t,w] = 1;
}
```

simply converts the solution of the discrete resource problem into a solution to the original problem. It considers all tasks in the order given by their starting dates and assigns to them the first worker available by assigning the appropriate element of **use** to one. Since the discrete capacity constraints are satisfied, a worker must be available at this date. Note that the instruction

```
forall(t in Tasks ordered by increasing dmin(a[t].start))
    tryall(w in Workers)
        use[t,w] = 1;
```

is enclosed in a **once** instruction, since only one such solution is required.

11.4.5 Search Procedures

This section reviews the search support provided by OPL for scheduling applications involving discrete resources. In scheduling applications with discrete resources, two activities that require the same resource may overlap in time. However, the total capacity required by the activities at any given time t may not exceed the capacity available at time t. As a consequence, a solution to these

11. Scheduling

```
enum Tasks
   { masonry,carpentry,plumbing,ceiling,roofing,painting,windows,facade,garden,moving };
int duration[Tasks] = [7,3,8,3,1,2,1,2,1,1];
enum Workers { Thomas, Maite, Antoine };

scheduleHorizon = sum(t in Tasks) duration[t];
DiscreteResource pool(3);
Activity a[t in Tasks](duration[t]);
UnaryResource worker[Workers];
var int use[Tasks,Workers] in 0..1;

minimize
   a[moving].end
subject to {
   a[masonry] precedes a[carpentry]; a[masonry] precedes a[plumbing];
   a[masonry] precedes a[ceiling]; a[carpentry] precedes a[roofing];
   a[ceiling] precedes a[painting]; a[roofing] precedes a[windows];
   a[roofing] precedes a[facade]; a[plumbing] precedes a[facade];
   a[roofing] precedes a[garden]; a[plumbing] precedes a[garden];
   a[windows] precedes a[moving]; a[facade] precedes a[moving];
   a[garden] precedes a[moving]; a[painting] precedes a[moving];

   forall(t in Tasks)
      a[t] requires(1) pool;

   forall(t in Tasks, w in Workers)
      a[t] requires(use[t,w]) worker[w];

   forall(t in Tasks)
      sum(w in Workers) use[t,w] = 1;
};
search {
   setTimes(a);
   once {
      forall(t in Tasks ordered by increasing dmin(a[t].start))
         tryall(w in Workers)
            use[t,w] = 1;
   }
};
```

Statement 11.9: Converting Discrete to Unary Resources (house1.mod).

scheduling problems must assign specific times to the activities, producing a computational model fundamentally different from unary resources.

OPL also provides an instruction `setTimes(a)` that assigns starting dates to all activities in an array `a`. This instruction should not be applied blindly: in some situations it can miss solutions (e.g., when there are negative distance constraints of the form `a.start >= b.end - 3`). However, for problems with discrete resources, positive distance constraints, and activities with fixed duration, it may improve efficiency considerably over a more naive strategy.[2]

Under this hypothesis, the basic idea behind `setTimes` can be explained as follows. If `d` is the earliest date at which an activity can start, OPL selects a task that can be scheduled at date `d`. On backtracking, another task is selected to start at `d`, or execution fails if none are available. Tasks that are not scheduled at their earliest date are said to be *postponed*, and remain so until their starting dates are updated. Because of the nature of the problem and because of the properties of constraint propagation in OPL, one of these activities must start at date `d` in an optimal solution.

Once such a task is chosen, OPL considers the next date `n >= d` at which a non-postponed activity can start and also all the tasks that can start at `n`. Once again, one of these tasks is chosen nondeterministically to start at date `n`. Note that some tasks may have a minimal starting date smaller than `n`. These are of course the *postponed* activities, because there exists an optimal solution in which these tasks do not take a value in the range `d+1..n`. If the current partial solution can be extended to an optimal solution, the starting dates of these tasks have to be updated, enabling OPL to reconsider them. Moreover, if the latest starting date of one of the postponed activities is smaller than `n`, then this postponed activity cannot have its starting date updated, meaning that the current partial solution cannot be extended to an optimal solution. This process is iterated until all the tasks have been assigned a starting date (a solution has been found) or only postponed activities remain (failure).

Finally, since it may be important in some applications to choose which resources and activities should be considered next, OPL offers, for discrete resources, generalizations of the functions `localSlack` and `globalSlack`, presented earlier for unary resources. As mentioned previously, these functions give an approximate measure of the tightness of a resource.

11.5 Reservoirs

Unary and discrete resources are appropriate modeling tools when activities only require (or consume) resources. Some applications, however, may have a combination of activities, some requiring resources and others providing resources. For instance, an activity may require a resource `plumber`, while the plumber `Joe` may provide the resource `plumber`. Resources that can be required and provided are called *reservoirs* in OPL.

---

[2] Of course, it is often possible to make choices to come into a position where `setTimes` can be applied naturally.

11.5.1 Summary of the Concepts

A reservoir is declared as in

```
Reservoir plumbing(3);
```

which declares a reservoir `plumbing` of maximum capacity 3 and initial capacity 0. The capacity of the reservoir specifies the difference between the supply (i.e., what is provided or produced) and the demand (what is required or consumed). A second parameter can be added to specify the initial capacity as in

```
Reservoir plumbing(3,1);
```

As usual, it is also possible in OPL to declare arrays of reservoirs. Activities can require reservoirs in the same way as they require discrete resources:

```
a requires(2) plumbing
```

specifies that activity `a` requires two units of plumbing during its execution. Activities can also consume the reservoir, as in

```
a consumes(2) plumbing
```

In addition to these constraints, activities can also provide and produce reservoirs. The constraint

```
b provides(2) plumbing
```

specifies that activity `b` provides two units of plumbing during its execution, while the constraint

```
b produces(2) plumbing
```

specifies that activity `b` produces two units of plumbing from its end date to the horizon. Of course, providing and producing are the counterparts to requiring and consuming.

11.5.1.1 The House Problem with Reservoirs

We now illustrate reservoirs on an application that brings up many modeling issues. The problem consists, once again, of building a house and the activities are the same as previously described. However, this time, the starting dates of these activities are known and the issue is to allocate a "worker" to each of them. Each of the activities can be performed by a group of workers. In addition, each group of workers has a maximum capacity, so that only a subset of its workers can be allocated at any given time. Figure 11.5 gives some of the instance data for this problem: the name of each task, its duration, its starting date, and the groups that can perform the task. There are three groups, `g1`, `g2` and `g3`. Group `g1` has three workers, `Thomas`, `Brett`, and `Matthew`, and

Name	Duration	Start	Groups
masonry	7	0	{g1, g2}
carpentry	3	7	{g1, g3}
plumbing	8	7	{g2}
ceiling	3	7	{g1, g3}
roofing	1	10	{g1, g3}
painting	2	10	{g2, g3}
windows	1	11	{g1, g3}
facade	2	15	{g1, g2}
garden	1	15	{g1, g2, g3}
moving	1	17	{g1, g3}

Figure 11.5: Instance Data for the House Problem with Workers.

has a capacity of 2. Group `g2` has one worker, `Scott`, and group `g3` has also one worker, `Bill`. The goal of the application is to allocate workers to tasks while minimizing the time spent by a worker on the construction site.

The model for this application, given in Statement 11.10, is non-trivial and involves a number of previously seen concepts. The first set of declarations consists primarily of associating an activity with each task. Of course, the duration and the starting date of each activity are well known: the duration is specified during the activity declaration, while the starting date is enforced by the constraint

```
forall(t in Tasks)
    task[t].start = start[t];
```

The second set of declarations concerns the groups of workers. The key idea in this model is to associate a reservoir with each group. The activities require units from these reservoirs, while the workers provide units for these reservoirs. In addition to the reservoirs, the model uses a two-dimensional array `perform` of Boolean variables to keep track of which group perform each activities. More precisely, `perform[g,t]` is 1 if group `g` performs activity `t` and 0 otherwise. Several constraints are expressed in terms of these data. The constraint

```
forall(t in Tasks)
    sum(g in mayperform[t]) perform[g,t] = 1;
```

specifies that each task must be performed by exactly one group. The constraint

```
forall(t in Tasks)
    forall(g in mayperform[t])
        task[t] requires(perform[g,t]) group[g];
```

11. Scheduling

```
enum Tasks
    {masonry,carpentry,plumbing,ceiling,roofing,painting,windows,facade,garden,moving};
int duration[Tasks] = [7,3,8,3,1,2,1,2,1,1];
int totalDuration = sum(t in Tasks) duration[t];
scheduleHorizon = totalDuration;
int start[Tasks] = [0,7,7,7,10,10,11,15,15,17];
Activity task[t in Tasks](duration[t]);

enum Groups { g1, g2, g3 };
int capacity[Groups] = [2, 1, 1];
{Groups} mayperform[Tasks] = #[
    masonry:{g1,g2}, carpentry:{g1,g3}, plumbing:{g2}, ceiling:{g1,g3}, roofing:{g1,g3},
    painting:{g2,g3}, windows:{g1,g3}, facade:{g1,g2}, garden:{g1,g2,g3}, moving:{g1,g3}]#;
var int perform[Groups,Tasks] in 0..1;
Reservoir group[g in Groups](capacity[g]);

enum Workers { Thomas, Brett, Matthew, Scott, Bill };
{Workers} workers[Groups] = #[g1:{ Thomas, Brett, Matthew }, g2:{ Scott }, g3:{ Bill }]#;
var int durationWorkers[Workers] in 0..totalDuration;
Activity worker[t in Workers](durationWorkers[t]);

minimize max(w in Workers) durationWorkers[w]
subject to {
    forall(t in Tasks) task[t].start = start[t]; /* starts of Tasks */
    forall(t in Tasks) /* resource constraints */
        forall(g in mayperform[t])
            task[t] requires(perform[g,t]) group[g];
    forall(t in Tasks) sum(g in mayperform[t]) perform[g,t] = 1; /* one resource must be used */
    forall(g in Groups) /* providing the resources */
        forall(w in workers[g])
            worker[w] provides group[g];
    forall(g in Groups) /* removing Symmetries */
        forall(ordered v, w in workers[g])
            worker[w].start >= worker[v].start & worker[w].end >= worker[v].end;
};
search {
    forall(t in Workers) {
        generate(durationWorkers[t]);
        generate(worker[t].start);
    };
    generate(perform); };
```

Statement 11.10: Allocating Workers for the House Problem (house4.mod).

specifies that a task `t` requires a group `g` if and only if `g` performs `t`. This is expressed concisely by using variable `perform[g,t]` as the capacity required. When `g` does not perform `t`, there is no requirement constraint. When `g` performs `t`, there is a requirement constraint of capacity 1.

The last set of declarations concerns the workers. Each worker is an activity whose duration is unknown, i.e., it is a variable ranging over `0..totalDuration`. These activities provide units for the groups in the constraint

```
forall(g in Groups)
    forall(w in workers[g])
        worker[w] provides group[g];
```

The model is now complete. The remaining instructions are included only to improve efficiency. The constraint

```
forall(g in Groups)
    forall(ordered v, w in workers[g])
        worker[w].start >= worker[v].start & worker[w].end >= worker[v].end;
```

removes symmetries between the workers to avoid exploring "equivalent" solutions, since workers are identical resources in this model. The choice specification

```
search {
    forall(t in Workers) {
        generate(durationWorkers[t]);
        generate(worker[t].start);
    };
};
```

simply defines the order of enumeration for the variables: considers the various workers and assigns a duration and a starting date to each of them in sequence.

11.6 Alternative Resources

Section 11.4.4 discussed how to use a discrete resource instead of a set of "equivalent" unary resources. It also showed how to transform the solution of the discrete resource problem into a solution to the original problem. In some applications, however, the unary resources may not be equivalent: for instance, unary resources may correspond to workers and only some workers may perform some of the tasks. But some of these unary resources may be "equivalent" for some activities in the problem. In these case, it is of course possible to use 0/1 variables, as in the previous section, to model the fact that an activity requires a resource from a pool of available resources. *Alternative resources* are another tool provided in OPL to model these applications.

11.6.1 Summary of the Concepts

Alternative resources are in fact sets of unary resources that can be declared, as in

```
UnaryResources worker[Workers];
AlternativeResources s(worker);
```

assuming that `worker` is an array of unary resources. This declaration simply specifies that `s` is a set of unary resources. This set can be used in `requires` constraints, as in

```
a requires s;
```

where `a` is an activity. This constraint specifies that activity `a` requires one of the unary resources in `s`. In addition to the `requires` constraints, OPL also provides an interesting tool to manipulate alternative resources: the constraint `activityHasSelectedResource(a,s,u)`, which holds if activity `a` has selected resource u in alternative resource `s`. Of course, this constraint can be negated and used as a higher-order constraint to express a variety of concepts.

11.6.2 The House Problem with Alternative Resources

To illustrate alternative resources, we consider the house problem one last time. Here the problem has the same set of tasks and associated durations as in the previous section. However, now each task must be performed by a worker from the set {joe, jim, jack}. As before, a worker cannot perform two tasks at the same time. In addition, some tasks can be performed by only some of the workers. The problem consists of assigning a starting date and a worker to all tasks so as to minimize the attendance of the workers on the site.

Statement 11.11 gives a model for this problem. The first set of declarations specifies the activities associated with the tasks, and the second set specifies the resources. Each worker is viewed as a unary resource and `worker` is an array of these resources. In addition, the model defines an alternative resource

```
AlternativeResources s(worker);
```

containing all workers. Note also that the model specifies the workers who cannot perform a given task. The last set of declarations again concerns the workers: it associates an activity `attendance[w]` with each worker `w`, which represents the attendance of the worker at the construction site. The duration of these activities is a variable ranging over the schedule time interval.

The problem constraints first express the precedence constraints and the fact that each task requires a worker:

```
forall(t in Tasks) task[t] requires s;
```

The next constraint

```
enum Tasks
   {masonry,carpentry,plumbing,ceiling,roofing,painting,windows,facade,garden,moving};
int duration[Tasks] = [7,3,8,3,1,2,1,2,1,1];
int totalDuration = sum(t in Tasks) duration[t];
scheduleHorizon = totalDuration;
Activity task[t in Tasks](duration[t]);

enum Workers { joe, jack, jim };
{Workers} cannotperform[Tasks] = #[
   masonry:{jim},carpentry:{jack},plumbing:{joe,jim},ceiling:{jack}, roofing:{jack},
   painting:{joe},windows:{jack},facade:{jim}, garden:{},moving:{jack}]#;
UnaryResource worker[Workers];
AlternativeResources s(worker);

var int durationWorkers[Workers] in 0..totalDuration;
Activity attendance[w in Workers](durationWorkers[w]);

minimize
   max(w in Workers) durationWorkers[w]
subject to {
   task[masonry] precedes task[carpentry]; task[masonry] precedes task[plumbing];
   task[masonry] precedes task[ceiling]; task[carpentry] precedes task[roofing];
   task[ceiling] precedes task[painting]; task[roofing] precedes task[windows];
   task[roofing] precedes task[facade]; task[plumbing] precedes task[facade];
   task[roofing] precedes task[garden]; task[plumbing] precedes task[garden];
   task[windows] precedes task[moving]; task[facade] precedes task[moving];
   task[garden] precedes task[moving]; task[painting] precedes task[moving];

   forall(t in Tasks) task[t] requires s;

   forall(t in Tasks)
      forall(w in cannotperform[t])
         not activityHasSelectedResource(task[t],s,worker[w]);

   forall(t in Tasks)
      forall(w in Workers)
         activityHasSelectedResource(task[t],s,worker[w]) =>
            attendance[w].start <= task[t].start & attendance[w].end >= task[t].end;
};
```

Statement 11.11: Allocating Workers for the House Problem with Alternative Resources (house3.mod).

```
forall(t in Tasks)
   forall(w in cannotperform[t])
      not activityHasSelectedResource(task[t],s,worker[w]);
```

removes from consideration the workers that cannot perform a task. More precisely, if `t` is a task and `w` is a worker who cannot perform the task, the constraint states that the resource `worker[w]` cannot be selected as a resource in the set `s` for activity `task[t]`. Finally, the last constraint

```
forall(t in Tasks)
   forall(w in Workers)
      activityHasSelectedResource(task[t],s,worker[w]) =>
         attendance[w].start <= task[t].start & attendance[w].end >= task[t].end;
```

expresses that a worker must be on the construction site to perform an activity. More precisely, whenever an activity `task[t]` has selected resource `worker[w]` from `s`, the worker must be present on the construction site before the activity starts and must remain on site at least until the activity is completed. Note that the OPL statement is quite close to the informal description.

The statement could be improved by using the search procedure

```
search {
   assignAlternatives;
   setTimes;
};
```

Instruction `assignAlternatives` is a nondeterministic instruction that, when it succeeds, guarantees that a resource has been selected for each constraint using an alternative resource. The rest of the search procedure specifies how to solve the problem when all resources are allocated. Note also that constraints such as `activityHasSelectedResource(a,s,u)` can be used to define a strategy tailored to the problem at hand.

11.7 Notes and References

The bridge problem came originally from [2] and its first constraint programming solution was described in [28]; the ship-loading program came originally from [22] and its first constraint programming solution appeared in [1]. As mentioned, that paper also contained the first solution to the perfect square problem, on which the model presented in this chapter was based. The discussion of Section 11.4.4 is based on a similar discussion in [24]. The various versions of the house problem are taken from [24].

Bibliography

[1] A. Aggoun and N. Beldiceanu. Extending CHIP to Solve Complex Scheduling and Packing Problems. In *Journées Francophones de Programmation Logique*, Lille, France, 1992.

[2] M. Bartusch. *Optimierung von Netzplaenen mit Anordnungsbeziehungen bei Knappen Betriebsmitteln*. PhD thesis, Fakultaet fuer Mathematik und Informatik, Universitaet Passau, Germany, 1983.

[3] N. Christofides. *Graph Theory: An Algorithmic Approach*. Academic Press, New York, 1975.

[4] A. Colmerauer. An Introduction to Prolog III. *Commun. ACM*, 28(4):412–418, 1990.

[5] A. Colmerauer, H. Kanoui, and M. Van Caneghem. Prolog, Bases Théoriques et Développements Actuels. *T.S.I. (Techniques et Sciences Informatiques)*, 2(4):271–311, 1983.

[6] G.B. Dantzig. *Linear Programming and Extensions*. Princeton University Press, Princeton, N.J., 1963.

[7] M. Dincbas, H. Simonis, and P. Van Hentenryck. Solving the Car Sequencing Problem in Constraint Logic Programming. In *European Conference on Artificial Intelligence (ECAI-88)*, Munich, Germany, August 1988.

[8] M. Dincbas, P. Van Hentenryck, H. Simonis, A. Aggoun, T. Graf, and F. Berthier. The Constraint Logic Programming Language CHIP. In *Proceedings of the International Conference on Fifth Generation Computer Systems*, Tokyo, Japan, December 1988.

[9] R. Fourer, D. Gay, and B.W. Kernighan. *AMPL: A Modeling Language for Mathematical Programming*. The Scientific Press, San Francisco, CA, 1993.

[10] M.R. Garey and D.S. Johnson. *Computers and Intractability*. W.H. Freeman and Company, New York, 1979.

[11] R.S Garfinkel and G.L Nemhauser. *Integer Programming*. John Wiley & Sons, New York, 1972.

[12] D. Gusfield and R.W. Irving. *The Stable Marriage Problem: Structure and Algorithms*. The MIT Press, Cambridge, MA, 1989.

[13] J. Jaffar and M. Maher. Constraint Logic Programming: A Survey. *Journal of Logic Programming*, 19/20:503–582, May/July 1994.

[14] J. Jaffar, S. Michaylov, P.J. Stuckey, and R. Yap. The CLP($\Re$) Language and System. *ACM Trans. on Programming Languages and Systems*, 14(3):339–395, 1992.

[15] M.J. Maher. Logic Semantics for a Class of Committed-Choice Programs. In *Fourth International Conference on Logic Programming*, pages 858–876, Melbourne, Australia, May 1987.

[16] K. McAloon and C. Tretkoff. 2LP: Linear Programming and Logic Programming. In V. Saraswat and P. Van Hentenryck, editors, *Principles and Practice of Constraint Programming*. The MIT Press, Cambridge, Ma, 1995.

[17] W. Older and A. Vellino. Extending Prolog with Constraint Arithmetics on Real Intervals. In *Canadian Conference on Computer & Electrical Engineering*, Ottawa, 1990.

[18] C.H. Papadimitriou and K. Steiglitz. *Combinatorial Optimization: Algorithms and Complexity*. Prentice-Hall, Englewood Cliffs, NJ, 1982.

[19] B.D. Parrello. CAR WARS: The (Almost) Birth of an Expert System. *AI Expert*, 3(1):60–64, January 1988.

[20] J-F. Puget. A C++ Implementation of CLP. In *Proceedings of SPICIS'94*, Singapore, November 1994.

[21] J-F. Puget and M. Leconte. Beyond the Glass Box: Constraints as Objects. In *Proceedings of the International Symposium on Logic Programming (ILPS-95)*, Portland, OR, November 1995.

[22] Roseaux. *Programmation linéaire et extensions; problèmes classiques*, volume 3 of *Exercices et problèmes résolus de Recherche Opérationnelle*. Masson, Paris, 1985.

[23] Ilog SA. Ilog Planner 2.0 Reference Manual, 1997.

[24] Ilog SA. Ilog Scheduler 2.0 Reference Manual, 1997.

[25] Ilog SA. Ilog Solver 4.0 Reference Manual, 1997.

[26] V.A. Saraswat. *Concurrent Constraint Programming Languages*. PhD thesis, Carnegie-Mellon University, 1989.

[27] G. Smolka. The Oz Programming Model. In Jan van Leeuwen, editor, *Computer Science Today*. LNCS, No. 1000, Springer Verlag, 1995.

[28] P. Van Hentenryck. *Constraint Satisfaction in Logic Programming*. Logic Programming Series, The MIT Press, Cambridge, MA, 1989.

[29] P. Van Hentenryck. Constraint Logic Programming. *Knowledge Engineering Review*, 6(3), 1991.

[30] P. Van Hentenryck. Scheduling and Packing in the Constraint Language cc(FD). In *Intelligent Scheduling*. M. Zweben and M. Fox (Eds.), Morgan Kaufmann, 1994.

[31] P. Van Hentenryck. *Encyclopedia of Science and Technology*, chapter Constraint Programming. Marcel Dekker, 1997.

[32] P. Van Hentenryck and J-P. Carillon. Generality Versus Specificity: An Experience with AI and OR Techniques. In *Proceedings of the American Association for Artificial Intelligence (AAAI-88)*, (St. Paul, MN), August 1988. AAAI, Menlo Park, Calif.

[33] P. Van Hentenryck and Y. Deville. The Cardinality Operator: A New Logical Connective and its Application to Constraint Logic Programming. In *Eighth International Conference on Logic Programming (ICLP-91)*, Paris (France), June 1991.

[34] P. Van Hentenryck and V. Saraswat. Strategic Directions in Constraint Programming. *ACM Computing Surveys*, 28(4), December 1996.

[35] P. Van Hentenryck, V. Saraswat, and Y. Deville. The Design, Implementation, and Evaluation of the Constraint Language cc(FD). In *Constraint Programming: Basics and Trends*. Springer Verlag, 1995.

[36] W. Winston. *Operations Research: Applications and Algorithms*. PWS Publishing, 1996.

Index

`::`, 79
`AlternativeResources`, 75
`DiscreteResource`, 74
`Reservoir`, 75
`UnaryResource`, 74
`abs`, 86, 87
`activityHasSelectedResource`, 106, 137
`activity`, 72
`all`, 102
`alldifferent`, 101
`assert`, 76
`assignAlternative`, 137
`bound`, 93, 129
`branchLow`, 135, 166
`branchUp`, 135, 166
`branch`, 135, 166
`breakable`, 73
`break`, 103
`breakOnDuration`, 103
`capacityMax`, 105
`capacityMin`, 105
`card`, 87
`ceil`, 87
`circuit`, 101
`circuit`, 191
`constraint`, 76
`consumes`, 104
`diff`, 90
`disjunctive`, 104
`display`, 139
`distToInt`, 87
`distribute`, 40, 102
`dmax`, 92
`dmin`, 92
`dnexthigher`, 92
`dnextlower`, 92
`dsize`, 92
`edge-finder`, 104
`enum`, 58
`first`, 87
`float`, 58
`floor`, 87
`forall`, 125
`frac`, 87
`ftoi`, 87
`generateMax`, 135
`generateMin`, 135
`generateSeq`, 135
`generateSize`, 135
`generate`, 135
`globalSlack`, 92
`if-then-else`, 108, 128
`infinity`, 58, 87
`initialize`, 62

`inter`, 90
`int`, 57
`in`, 94
`isPossibleFirst`, 92
`isPossibleLast`, 92
`isRanked`, 92
`last`, 87
`let`, 130
`localSlack`, 92
`maximize`, 106
`maxint`, 57, 86
`max`, 88
`minimize`, 106
`min`, 88
`mod`, 86
`nbOccur`, 92
`nbPossibleFirst`, 92
`nbPossibleLast`, 92
`nearest`, 87
`nextc`, 87
`next`, 87
`not`, 94
`onDomain`, 101, 134
`onRange`, 101, 134
`onValue`, 101, 133
`ordered`, 113
`ord`, 87
`periodicBreak`, 103
`piecewise`, 89
`prevc`, 87
`prev`, 87
`produces`, 105
`prod`, 88
`provides`, 105
`range`, 59
`rankGlobal`, 136
`rankLocal`, 136
`rank`, 136
`reducedCost`, 92
`regretdmax`, 92
`regretdmin`, 92
`requires`, 103, 104
`scheduleHorizon`, 72
`scheduleOrigin`, 72
`search`, 121
`select`, 129
`setTimes`, 136
`simplexValue`, 92
`solve`, 106
`splitLow`, 135, 166
`splitUp`, 135, 166
`split`, 166
`sqrt`, 87
`struct`, 63

```
subset, 94
sum, 88
symdiff, 90
trunc, 87
tryRankFirst, 136
tryRankLast, 136
tryall, 124
try, 121
union, 90
using, 104
var, 68
when, 132
while, 129
with linear relaxation, 106, 181
```

2LP, 4

a

activity, 51, 72
  breakable, 73
aggregate
  array, 102
  operator, 15, 88
aggregating
  results, 141
algebraic, 2
alternative resource, 75, 105, 137, 232
AMPL, 1, 2, 8, 176
array, 14, 60
  initialization, 61

b

backtracking, 33, 35, 123, 130
blending, 28, 148
block, 107
BNR-Prolog, 4
Boolean connective, 91
branch and bound, 165–167
branching, 135
  instruction, 166
breakable activity, 73
breakpoint, 88
break, 73, 103
bridge, 205

c

capacity, 51, 105
car sequencing, 184
cardinality constraint, 99
case-sensitive, 55
casting, 87
cc(fd), 4
CHIP, 4, 53
choice, 33
CLP(R), 4
command language, 2, 8
complexity, 173
computational model, 35
computed initialization, 79
concave, 175
concurrent constraint programming, 4
conditional statement, 108, 128
conjunction, 91, 94
constraint, 97, 132
  basic, 98
  cardinality, 99
  cumulative, 53
  declaration, 75
  discrete, 98
  entailment, 4
  higher-order, 39, 99
  name, 108
  redundant, 40, 182
  surrogate, 40
constraint logic programming, 4
constraint programming, 3, 31
constraint satisfaction, 126
constraint store, 4, 106, 121, 132
constraint-driven
  computation model, 4
  construct, 8, 132
consume, 51
convention, 9
convertion
  from discrete to unary resources, 224
convex, 175
cumulative constraint, 53

d

data
  consistency, 76
  declaration, 15
  structure, 59
data-driven construct, 132
decision problem, 3
declarative, 5

Index

default strategy, 33
derived result, 142
discrete
  constraint, 98
  resource, 51, 74, 104, 136, 218, 224
disjunction, 47, 91, 94
display, 23
  instruction, 139
  tuples, 142
domain, 32
duration, 51
dynamic ordering, 125–126

e

edge finder, 4, 103
entailment, 4
enumerated
  expression, 87
  type, 58
equivalence, 91, 94
Euler tour, 191
expression, 85

f

failure, 38, 121
field, 22, 64
file initialization, 82
filtering, 114–115, 141
first-fail principle, 35
fixed-charge, 163
float, 58
  constraint, 97
  expression, 87
floating-point number, 55
flow algorithms, 4
frequency allocation, 194

g

GAMS, 1, 2
generate, 33
generation
  of values, 35, 134
generic
  array, 63
  initialization, 41
global
  constraint, 40, 101
  slack, 136
ground, 108

h

Hamiltonian circuit, 102
heuristic, 38, 121, 179, 194
higher-order constraint, 39, 99
horizon, 72
house problem, 224

i

identifier, 55, 85
Ilog Solver, 4
implication, 91, 94, 100
index set of arrays, 15, 60
infeasibility, 174
initialization, 17
  array, 61
  computed, 80
  generic, 41
  inline, 77
  file, 82
  offline, 79
  record, 64
integer, 55, 57
  division, 86
  expression, 86
integer programming, 2, 3, 13, 25, 135, 157
inventory model, 166
isolating data, 16

j

job-shop scheduling, 214

k

knapsack, 25

l

large-scale, 115, 154
linear programming, 2, 14, 13, 145
linear relaxation, 98, 106, 135, 165–166, 181
local
  consistency, 98
  slack, 134
logical combination of constraint, 98

m

magic series, 39, 53
map coloring, 42
mathematical programming, 1–2
membership, 94, 191
MILP, 28
misconceptions, 5
mixed integer-linear programming, 28, 166
modeling language, 1
model, 56
multi-directional, 100
multi-knapsack, 25
multi-period production planning, 145
multicommodity flow, 153

n

negation, 91, 94
NP-complete, 3
nondeterminism, 4, 8, 33, 35, 49, 121
nonlinear programming, 3
nonterminal, 55

o

offline initialization, 77
operator precedence, 94
optimization problem, 13
ordering, 129
  dynamic, 125–126
  value, 126
  variable, 126
origin, 72
Oz, 4

p

parenthesis, 94
perfect square, 47, 53, 223
periodic, 103
piecewise linear
  function, 88, 175
  programming, 168
precedence, 51, 103
preprocessing, 27, 157
production planning, 14, 18, 145
project, 16
Prolog II, 4
Prolog III, 4
propagation, 104
pruning, 32

q

quantifier, 15, 125
queens problem, 34
queries, 66

r

rack configuration, 198
range, 28, 59
record, 18, 63
  initialization, 64
reduced cost, 93
redundant, 49
  constraint, 40, 184
reflective function, 92, 108
regret, 177
relation, 85, 92
  relation in expression, 91
relaxation, 121
reservoir, 105, 228
resource, 73, 103
  allocation, 2

s

satisfiability, 4
scheduling, 2, 4, 50, 70, 102, 136
scope, 64, 127
search, 4, 5, 8
  procedure, 2, 34, 38, 47, 121
  space, 40
separation
  between model and data, 2
  separation of constraints and search, 5
set, 65
  covering, 157
  expression, 90
ship loading, 220
slope, 89
sparse, 67
sparsity, 115, 153
stable marriage, 43, 53
starting date, 51
OPL studio, 16
surrogate constraint, 40
syntactic conventions, 55

Index

t

terminal, 55
tightness, 136
transportation, 153
tuples of parameters, 113

u

unary resource, 74, 103, 136, 205, 224
universal quantifier, 16, 107

v

value ordering, 126
variable, 68
  as index, 43, 99, 179
  ordering, 126
  ranging an enumerated set, 42, 43
  0-1, 39, 99

w

warehouse location, 159, 177